Lecture Notes in Physics

T0225024

The Lecture Notes in Physics

The series Lecture Notes in Physics (LNP), founded in 1969, reports new developments in physics research and teaching – quickly and informally, but with a high quality and the explicit aim to summarize and communicate current knowledge in an accessible way. Books published in this series are conceived as bridging material between advanced graduate textbooks and the forefront of research and to serve three purposes:

- to be a compact and modern up-to-date source of reference on a well-defined topic
- to serve as an accessible introduction to the field to postgraduate students and nonspecialist researchers from related areas
- to be a source of advanced teaching material for specialized seminars, courses and schools

Both monographs and multi-author volumes will be considered for publication. Edited volumes should, however, consist of a very limited number of contributions only. Proceedings will not be considered for LNP.

Volumes published in LNP are disseminated both in print and in electronic formats, the electronic archive being available at springerlink.com. The series content is indexed, abstracted and referenced by many abstracting and information services, bibliographic networks, subscription agencies, library networks, and consortia.

Proposals should be sent to a member of the Editorial Board, or directly to the managing editor at Springer:

Christian Caron
Springer Heidelberg
Physics Editorial Department I
Tiergartenstrasse 17
69121 Heidelberg / Germany
christian.caron@springer.com

P. Bydžovský
A. Gal
J. Mareš (Eds.)

Topics in Strangeness
Nuclear Physics

 Springer

Editors

Petr Bydžovský
Jiří Mareš
Nuclear Physics Institute
Academy of Sciences
of the Czech Republic
250 68 Rez near Prague, Czech Republic
bydz@ujf.cas.cz
mares@ujf.cas.cz

Avraham Gal
Racah Institute of Physics
The Hebrew University
91904 Jerusalem, Israel
avragal@vms.huji.ac.il

P. Bydžovský, A. Gal and J. Mareš (Eds.), *Topics in Strangeness Nuclear Physics*, Lect. Notes Phys. 724 (Springer, Berlin Heidelberg 2007), DOI 10.1007/ 978-3-540-72039-3

ISSN 0075-8450

ISBN 978-3-642-09122-3   e-ISBN 978-3-540-72039-3

Springer is a part of Springer Science+Business Media
springer.com
© Springer-Verlag Berlin Heidelberg 2007
Softcover reprint of the hardcover 1st edition 2007

Cover design: eStudio Calamar S.L., F. Steinen-Broo, Pau/Girona, Spain

Preface

The present book includes the edited versions of lectures presented at the XVIIIth Indian-Summer School of Physics held at Řež/Prague during the period 3–7 October, 2006 (http://rafael.ujf.cas.cz/school). The Indian-Summer School series started in 1988 and soon gained its firm position among European schools of intermediate-energy nuclear and hadronic physics. The School has always aimed at offering students an opportunity to gain access to contemporary theoretical and experimental achievements and to future prospects in selected fields from renowned scholars. The XVIIIth School was devoted to strangeness nuclear physics.

Strangeness nuclear physics bears a broad impact on contemporary physics since it lies at the intersection of nuclear and elementary particle physics and, furthermore, has significant implications for astrophysics. An hyperon embedded in the nuclear medium presents a unique probe of the deep nuclear interior which makes it possible to study a variety of otherwise inaccessible nuclear phenomena, and thereby test nuclear models. The added strange hadron, whether hyperon or antikaon, introduces an SU(3)-flavour dimension to traditional nuclear physics. It enables one to study directly models of baryon–baryon and meson–baryon interactions as well as effective field theory approaches which encode the basic ingredients of QCD at low energy. This was the main theme of the lectures given by Rob Timmermans. In this book, the subject is discussed comprehensively in the review of Johann Haidenbauer, Ulf Meißner, Andreas Nogga, and Henk Polinder who underline the modern effective field theory approach in its first application to the hyperon–nucleon interaction.

In-medium weak decays of hyperons serve as a unique tool for exploring novel aspects of the weak interactions, in particular by studying the non-mesonic decay modes induced by one nucleon in Λ hypernuclei: $\Lambda N \rightarrow NN$. On the experimental side, correlation measurements allow now to single out the effect of final state interaction and the effect of added two-nucleon induced processes which distort the measured final two-nucleon spectra. On the theoretical side, full account has been taken of the short-range nature of the weak interaction imposed by the exchange of heavy mesons. A long-standing discrepancy was resolved between the experimentally deduced ratio of neutron-to-proton induced weak decay rates and the theoretical prediction using hadronic degrees of freedom within a four-fermion weak-interaction formulation. This was the main theme of the lectures given by Assumpta Parreño.

There is a growing evidence for the significance of strange particles in astrophysics; hyperons (and possibly antikaons) appear to be important constituents of the central part of neutron stars, where strangeness is likely to be materialized macroscopically in dense hadronic matter. Whereas experimental progress has been substantial for single (strangeness $S = -1$) Λ hypernuclei, little is known about multistrange hypernuclei. In the $S = -2$ sector, the unambiguous determination of the $_{\Lambda\Lambda}^{6}\mathrm{He}$ double-Λ hypernucleus has revised one's notion on the $\Lambda\Lambda$ interaction which appears now to be considerably weaker than believed before, although definitely attractive on the whole. This has significant implications for the equation of state in compact stars. An equally important subsector of $S = -2$ hypernuclei consists of the study of Ξ hypernuclei on which there is hardly any experimental knowledge. Aspects of multistrangeness, $S = -2$ and beyond, as explored to date and as projected for the near future in forthcoming new experimental accelerator facilities, were highlighted in the lectures of Ed Hungerford and Tomofumi Nagae.

There is considerable interest lately in the strongly attractive as well as absorptive antikaon ($\bar{\mathrm{K}}$) nuclear interaction and whether or not it can lead to the existence of fairly narrow deeply bound states. This subject was reviewed in the lectures given by Avraham Gal, but since it was deemed by the editors more speculative than the other topics included in this book, it was decided to exclude it for the time being.

The evolution of strangeness nuclear physics, described thoroughly in Hungerford's and Nagae's lectures, started with the discovery of the first Λ hypernucleus in 1952. The experimental study of Λ hypernuclei in the first two decades uncovered major general properties of these new objects, in spite of being limited to ground states of light species by working almost exclusively with nuclear emulsions. During the late 1970s, (K^-, π^-) strangeness exchange counter experiments at the CERN PS and the BNL AGS revealed rich Λ hypernuclear spectra, dominated by coherent excitations, over a wide range of the periodic table, leading to the elucidation of in-medium Λ-nucleon interaction components. The late 1980s witnessed the beginning of (π^+, K^+) associated production experiments performed first at the BNL AGS and then dominantly and until very recently at the KEK PS. These experiments, with selectivity complementary to that of the (K^-, π^-) reaction, revealed unprecedented single-particle spectra in light, medium-weight, and heavy hypernuclei. The (π^+, K^+) reaction provided also relatively high-resolution excitation spectra in light hypernuclei. Electromagnetic radiation and weak decays of Λ hypernuclear levels have been studied by coincidence measurements, tagging to the outgoing mesons in the primary (K^-, π^-) and (π^+, K^+) production experiments and to some of the decay products. In particular, the development of γ-ray spectroscopy in light Λ hypernuclei has yielded invaluable information on the spin dependence of the in-medium ΛN interaction. This subject was discussed thoroughly in the comprehensive lectures given by John Millener. Recently, the double-charge-exchange (π^-, K^+) associated production has also been used at KEK, first to map out the continuum spectrum of Σ hypernuclei, which with the exception of $^4_\Sigma$He appear then not to exist in the bound-state region, and second to produce for the first time in accelerator experiments a neutron-rich hypernucleus $^{10}_\Lambda$Li.

A new stage in strangeness nuclear physics started recently with the application of high-quality electron beams at the Thomas Jefferson Laboratory through the electromagnetic $(e, e'K^+)$ associated production reaction, and with new facilities in Frascati and Mainz. Further progress of strangeness nuclear physics can be expected in few years with the new facilities of J-PARC, a 50 GeV PS in Tokai, Japan, and of the PANDA detector system in the $p\bar{p}$ collider section of the FAIR project in Darmstadt, Germany. The study of $S = -2$ hypernuclei, Ξ as well as $\Lambda\Lambda$, is scheduled to be a major theme in these forthcoming facilities. These developments and the fact that the School took place just few days before the IX International Conference on Hypernuclear and Strange Particle Physics (HYP 2006) in Mainz provided a forceful motivation for the choice of this year's subject.

We are pleased that some of the most renowned scholars in the field of strangeness nuclear physics agreed to participate and thereby contribute significantly to the high scientific level of the School. We are most grateful to the lecturers for all their efforts in preparing, presenting, and finally writing up their excellent lectures which cover a broad range of topics. Thanks are also due to the students for the well-prepared, interesting seminars and their contributions to the discussions throughout the School. We are obliged to Daniel Gazda for his assistance in editing these Lecture Notes.

In addition to the Nuclear Physics Institute, Řež, and the Institute of Particle and Nuclear Physics of Charles University, Prague, the School was supported by the JINR-CR Collaboration Committee, the CERN-CR Collaboration Committee, the Votruba Blokhintsev Program for Theoretical Physics, and the company Optaglio Ltd. We thank each of these institutions for their generous support which enabled us to make the costs of the School accessible for students and young scientists. Last but not least, we would like to express special thanks to our secretary, Ms. Růžena Ortová, for running the School smoothly and effectively.

The number of participants in the School was a clear signature of the vitality of the field. In spite of its 'age' of over 50 years, strangeness nuclear physics has not lost its charm and beauty. On the contrary, it keeps up to its young spirit in addressing new challenging problems and questions in the quest for understanding the role of strangeness in the many facets envisaged by contemporary physics.

Prague and Jerusalem,
December 2006

Petr Bydžovský
Avraham Gal
Jiří Mareš

Contents

The Production of Strange Nuclear Systems

Ed V. Hungerford

Department of Physics, University of Houston, Houston, TX 77204, USA
hunger@uh.edu

1 A Perspective

It sometimes seems unfortunate that physics is taught from an axiomatic perspective. Which is to say that in the classroom, physical phenomena are derived from a set of axioms, leaving the impression that nature had no other choice in the matter. While this emphasizes mathematical beauty, it hardly does justice to the way physics actually proceeds to discover these postulates. The subject of strangeness in nuclear systems will be introduced with the encouragement that all students review this field from a historical perspective in order to obtain a clearer understanding of the inductive reasoning process that is crucial to the development of natural science.

In the late 1940's and early 1950's the concept of a strangeness quantum number was not understood [1]. "Strange" particles were observed when an emulsion was exposed to cosmic rays, and these particles seemed to be produced with high probability, yet had long lifetimes ($\approx 10^{-10}$ s). Thus one could infer that production and decay occurred by different mechanisms, but evidence that these particles were produced as pairs was mixed, so that explanations involving associated production, requiring a new, conserved quantum number were not generally accepted.

It was also at this time, 1952, that a serendipitous event, Fig. 1, was seen in a cosmic ray interaction with an emulsion nucleus [2]. The event resulted in a heavy remnant that traveled a long distance before it decayed, releasing substantial energy. At first, it was thought that the remnant contained a bound pion in a nucleus which eventually annihilated. However, when that explanation was found untenable, it was proposed that this remnant contained one of the new "strange" particles. This inductive leap may seem obvious now, but at that time it was a seminal proposal. It occurred because minds were prepared to accept the idea of another baryon having a new, conserved quantum number [3].

After the discovery of the first hypernuclear event, investigators began to explore this new field, although scanning of emulsion detectors was quite

E. V. Hungerford: *The Production of Strange Nuclear Systems*, Lect. Notes Phys. **724**, 1–29 (2007)
DOI 10.1007/978-3-540-72039-3_1

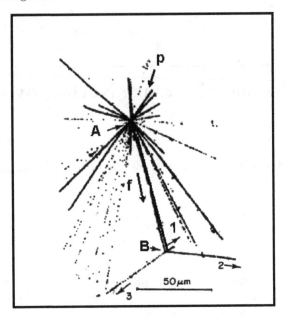

Fig. 1. The first hypernuclear event. Track P is the incident cosmic ray, which interacts at A, forming a hypernuclear track, f, decaying at B into tracks 1, 2, and 3 [2]

labor intensive, and results proceeded slowly. Approximately 20 years after the first observation, 22 different hypernuclear species, all essentially in the nuclear 1s and 1p shells, had been identified, their binding energies cataloged, and their decays observed [4]. Within the last 35 years or so, the use of electronic counters and modern accelerators has extended this list [6] to include heavier systems. As with normal nuclei, hypernuclear binding energies show saturation, as seen in Fig. 2 and Table 1.

A hypernucleus is constructed from a normal nucleus, atomic weight, A, and atomic number, Z, with the addition of one or more bound hyperons. A hyperon is a baryon composed of at least one strange, valence quark, i.e. a Λ, Σ, Ξ, or Ω^-. For example, the hypernucleus $^{12}_{\Lambda}$C has 12 baryons, with one of these being a Λ hyperon. It has atomic number 6, as noted by the label, C, although for a hypernucleus the atomic number is a measure of the charge and not necessarily the number of protons. In a simplistic single particle model, the various hyperons, neutrons, and protons are all considered distinguishable particles, so each is placed in independent, effective-potential wells in which the Pauli exclusion principle is applied. This is illustrated in Fig. 3, which is a cartoon that helps to conceptualize the structure of various hypernuclei, their excited states, and their production mechanisms. As shown in the figure, the hyperon occupies the lowest shell (1s) when the hypernucleus is in its ground state.

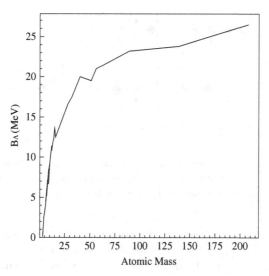

Fig. 2. The Λ binding energy as a function of atomic number, showing that the binding saturates

To conclude this introduction, it is the author's task to explore some of the more interesting aspects of the production of strange nuclear systems. Many problems of present interest were outlined in the early days of hypernuclear research, however, the ability to exploit these ideas resided in the available

Table 1. Experimental Λ binding energies of hypernuclei [4, 5, 6, 7]

Hypernucleus	B_Λ (MeV)	Hypernucleus	B_Λ (MeV)
$^{3}_{\Lambda}$H	0.13±0.05	$^{11}_{\Lambda}$B	10.24±0.05
$^{4}_{\Lambda}$H	2.04±0.04	$^{12}_{\Lambda}$B	11.37±0.06
$^{4}_{\Lambda}$He	2.39±0.03	$^{12}_{\Lambda}$C	10.76±0.19
$^{5}_{\Lambda}$He	3.12±0.02	$^{13}_{\Lambda}$C	11.69±0.12
$^{6}_{\Lambda}$He	4.18±0.10	$^{14}_{\Lambda}$C	12.17±0.33
$^{8}_{\Lambda}$He	7.16±0.70	$^{14}_{\Lambda}$N	12.17 †
$^{6}_{\Lambda}$Li	4.50 †	$^{15}_{\Lambda}$N	13.59±0.15
$^{7}_{\Lambda}$Li	5.58±0.03	$^{16}_{\Lambda}$O	12.42±0.05
$^{8}_{\Lambda}$Li	6.80±0.03	$^{28}_{\Lambda}$Si	16.6±0.2
$^{9}_{\Lambda}$Li	8.50±0.12	$^{32}_{\Lambda}$S	17.5±0.5
$^{7}_{\Lambda}$Be	5.16±0.08	$^{40}_{\Lambda}$Ca	20.0±0.5
$^{8}_{\Lambda}$Be	6.84±0.05	$^{51}_{\Lambda}$V	19.5 †
$^{9}_{\Lambda}$Be	6.71±0.04	$^{56}_{\Lambda}$Fe	21.0 †
$^{10}_{\Lambda}$Be	9.11±0.22	$^{89}_{\Lambda}$Y	23.1±0.5
$^{9}_{\Lambda}$B	8.29±0.18	$^{139}_{\Lambda}$La	24.5±1.2
$^{10}_{\Lambda}$B	8.89±0.12	$^{208}_{\Lambda}$Pb	26.3±0.8

† No error reported

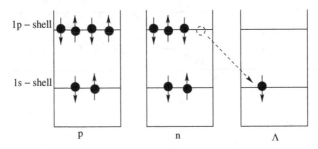

Fig. 3. A simplistic single particle model for the $^{12}_{\Lambda}$C hypernucleus, that helps to conceptualize hypernuclear structure. The figure shows a transition which changes a neutron into a Lambda, leaving the hypernucleus in its ground state

technology and the creativity of the researchers in the field. With the modern resurgence of interest in strange nuclear physics, new facilities and detection systems should provide steady, if not spectacular, growth in this field.

2 Formulation of the Hyperon Potentials

2.1 The Elementary Baryon-Baryon Potential

Although there is substantial information on the nucleon-nucleon interaction below pion threshold, there is little data and less analysis, of the hyperon-nucleon interaction. There are only some 600 or so low-momentum, Λ-N and Σ-N scattering events [8]. Even the Λ-N scattering lengths are not well known, and there is essentially no data on the Ξ-N, and Ω-N scattering, much less the hyperon-hyperon, interaction. The main reason for this, of course, is the short lifetime of the hyperons ($< 10^{-9}$ s), coupled with the additional complication that several hyperons are uncharged.

Thus one way to formulate a general baryon-baryon potential attempts to apply broken SU(3)$_f$ flavor symmetry to all the available baryon-baryon scattering data in order to take advantage of the much more copious N-N information [9, 10, 12, 13, 57]. Several groups have developed such potentials over the years, but probably the most used forms have been the various Nijmegen potentials. Table 2 compares the Λ-N singlet and triplet scattering lengths and effective ranges of several of these models, showing that the data cannot determine the scattering lengths.

The Λ has isospin 0, so Λ-N scattering occurs in only the isospin 1/2 state. On the other hand the Σ, has isospin 1 so that its interaction with a nucleon can occur in isospin states of 3/2 or 1/2. The Nijmegen potentials, and the limited data are consistent with significant Σ-N spin and isospin dependence, predicting strong attraction in the $^1S_0$, T = 3/2 and $^3S_1$, T = 1/2 channels and repulsion in the $^3S_1$, T = 3/2 and $^1S_0$, T = 1/2 channels. However, perhaps the most important difference between the Λ and Σ interactions is the strong

Table 2. Comparison of various model Λ-N amplitudes (in fm)

Model	Reference	a^s	r_0^s	a^t	r_0^t
Nijmegen D	[9]	−1.90	3.72	−1.96	3.24
Nijmegen F	[10]	−2.29	3.17	−1.88	3.36
Nijmegen SC89	[11]	−2.78	2.88	−1.41	3.11
Jülich A	[13]	−1.56	1.43	−1.59	3.16

conversion $\Sigma N \rightarrow \Lambda N$ with the release of some 80 MeV. This dominates the behavior of a Σ in the nuclear medium [14].

The inverse coupling $\Lambda N \rightarrow \Sigma N$ also affects the Λ-N interaction, requiring the inclusion of virtual Σ-N interactions. A schematic picture of this coupling is shown in Fig. 4. Since the Σ and Λ states are only 80 MeV apart as compared to the $\approx$ 300 MeV of the N and Δ isobars, three body forces and charge symmetry breaking are expected, and found to be more important in hypernuclei than in normal nuclear systems. Here one also notes that one pion exchange is prohibited by isospin conservation, so that the shorter ranged, two pion exchange process, along with heavier boson exchange, is assumed to account for the Λ-N interaction.

2.2 The Effective Λ-Nucleus Potential

When a Λ is embedded in a nuclear system, it interacts with all of the nucleons, and the resulting behavior is expressed as a nuclear many-body problem. The forces between baryons are hadronic, so the time scales of these interactions are some 10^{-23} s as compared to the Λ lifetime. Therefore the system is stable with respect to the strong interaction, and can be treated using well developed nuclear theory.

One then views a hypernucleus as a composite of baryons, each interacting through an effective potential generated by the other nucleons. In its most simplistic form a hypernucleus can be considered as a conventional nuclear core with a hyperon in a single particle state of the hyperon-nuclear effective potential. Here, we do not have to discuss the way this is done, which includes the superposition of single particle states and a diagonalization of the residual interactions. One can find more technical details in the references and in other contributions to these lectures [15, 16, 17].

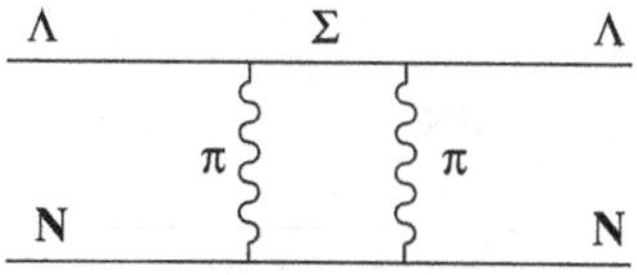

Fig. 4. A simple Feynman diagram showing Λ-Σ coupling within a hypernucleus leading to 3-body forces and charge symmetry breaking

Within a nucleus the general hyperon-nucleon potential can be expressed in the form;

$$V(r) = V_0 + V_s(\boldsymbol{S_N} \cdot \boldsymbol{S_Y}) + V_t S_{12} + V_{ls}(\boldsymbol{L} \times \boldsymbol{S^+}) + V_{als}(\boldsymbol{L} \times \boldsymbol{S^-})$$

In the above equation, $S_{12} = 3(\boldsymbol{\sigma_1} \cdot \hat{\boldsymbol{r}})(\boldsymbol{\sigma_2} \cdot \hat{\boldsymbol{r}}) - \boldsymbol{\sigma_1} \cdot \boldsymbol{\sigma_2}$ is the spin-tensor operator and $\boldsymbol{S^{\pm}} = 1/2(\boldsymbol{S_N} \pm \boldsymbol{S_Y})$ are the symmetric and anti-symmetric combinations of nucleon and hyperon spin operators. The potential contains both symmetric and anti-symmetric spin orbit terms, where the antisymmetric spin orbit term vanishes for the N-N potential due to the Pauli principle. Both of these terms are small in meson exchange potentials, and the spin orbit contribution to the effective Λ-N potential is nearly zero [5, 18]. This is not the case for the N-N potential.

2.3 An Example - the s-shell Hypernuclei

The s-shell hypernuclei illustrate many of the features of the Λ-N interaction, as discussed above. The level diagrams of all the s-shell hypernuclei are shown in Fig. 5. The hypertriton, $^3_\Lambda$H, with a binding energy of only 0.13 MeV, is the lightest, strange system. Thus one expects the Λ-N potential to be weaker

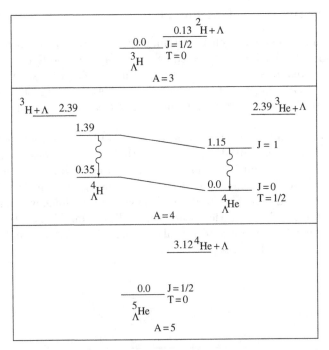

Fig. 5. Level diagrams of the s-shell hypernuclei showing the ground and excited states (energies are in MeV)

than for N-N. One also sees that the A = 4 system has two bound isobars, each with one excited state. To obtain the correct spin splitting, Coulomb energies, and charge-symmetry breaking involving Λ-Σ coupling must be included in a full many-body interaction calculation [19, 20]. It now remains to complete a simultaneous, systematic calculation of all the binding and excitation energies of the s-shell states, including many-body effects, and Λ-Σ coupling.

3 Production Mechanisms

3.1 The Distorted Wave Impulse Approximation

Obviously, one needs to bind a hyperon to a nucleus to produce a hypernucleus. Since the bound hyperon is in its own potential well, it has a maximum Fermi-momentum surface. This can be obtained from the binding energy of the least bound hypernuclear state and the uncertainty principle. To increase the likelihood that the hyperon binds to the nuclear medium, it should be produced with a momentum that is not much larger than this maximum Fermi-momentum. In addition as will be seen, kinematics also significantly affects the production reaction.

Figure 6 shows several quark flow diagrams which can be used to visualize various production processes. Although obvious, it should be noted that these diagrams are cartoons, and not intended to represent calculational procedures.

In general, a production reaction can be described by the distorted wave impulse approximation, DWIA [21]. This formulation views the target as a

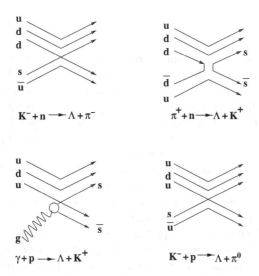

Fig. 6. Quark flow diagrams for several production mechanisms

collection of single particle levels, with a production amplitude occurring between the incident projectile and one of these particles. The other nucleons are then considered a set of spectator particles, contributing to an overall potential in which the interaction takes place. Thus the incident and exit particle waves are distorted by a nuclear optical potential. For example, the DWIA formulation for $\bar{K} + A \rightarrow \pi + {}_Y A$ is given by;

$$T_{AY} \propto \langle \xi_\pi^- \psi_Y | T | \xi_{\bar{K}}^+ \psi_A \rangle \qquad (1)$$

$$\propto \langle \pi\Lambda | t | \bar{K}N \rangle N_{\text{eff}}^{1/2} .$$

In this equation, the ξ are the distorted incident and final wave functions for the kaon and pion due to the nuclear optical potentials, and t is the elementary $\bar{K}N \rightarrow \pi Y$ transition matrix. The final form has been factored, removing the elementary amplitude from the total integral, so that the elementary transition amplitude, t, multiplies a density represented by the effective number of nucleons participating in the interaction. The above simple formulation can also be written to show specific spin and angular momentum dependence.

Distortions of the incident and exit waves generally do not change the shape of the angular distributions, but can reduce the reaction amplitudes by up to an order of magnitude. In addition, the factored elementary amplitude must be averaged over the Fermi momentum of the participating nucleons in the medium. This can reduce the cross section by 10–20%. Finally the DWIA approximation assumes that the reaction can be expressed by an elementary on-shell, t-matrix amplitude, although corrections for this approximation, and reaction processes that include the instantaneous interaction with more than one nucleon, are expected to be small.

3.2 Kinematics

The kinematics for several elementary reaction processes are shown in Fig. 7. As indicated in the figure, the $n(K^-, \pi^-)\Lambda$ reaction can have low, essentially zero, momentum transfer to a Λ or Σ hyperon. Thus the probability that this hyperon will interact with, or be bound to, the spectator nucleons is large. On the other hand reactions such as $n(\pi^+, K^+)\Lambda$ or $p(\gamma, K^+)\Lambda$ have high momentum transfer with respect to the Fermi momentum, and produce recoil hyperons that have a high probability of escaping the nucleus. Such reactions are sometimes loosely termed quasi-free processes (QF), although the hyperon actually experiences continuum, final-state interactions. In the case of higher momentum transfer, cross sections to bound states are reduced.

Finally, a K^- strongly interacts with nucleons through various resonant states. Thus incident kaons in a (K^-, π^-) reaction attenuate rapidly in nuclear matter, and the reaction strength peaks at the surface. In this case the (K^-, π^-) reaction most likely interacts with an outer shell neutron with little momentum transfer, simply replacing this neutron with a Λ in the same shell. On the other hand, energetic π^+ and K^+ particles have longer mean free paths

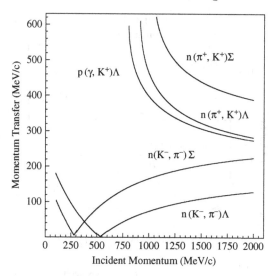

Fig. 7. The recoil momentum of a hyperon in various elementary reactions at 0° as a function of the incident particle momentum

in nuclear matter, and give larger momentum transfer to the hyperon. Thus they can interact with interior nucleons, and there can be significant angular momentum transfer. All of these features are illustrated in Fig. 8 which shows calculated spectroscopic factors for the formation of a hypernucleus by various reactions on an ^{56}Fe target [5]. One notes in particular the magnitude of the production strengths to various states and the QF strengths in each case.

3.3 Production into the Continuum

A QF process assumes that the DWIA reaction provides sufficient momentum to remove the hyperon from the nuclear medium, and by way of review, assumes that the reaction occurs through an elementary process on a nucleon moving in a potential well with Fermi momentum. In this model, the QF continuum spectrum can be obtained by calculating the statistical density of states for the reaction on a single-particle nuclear state as taken from a Fermi-gas distribution of nucleons. This produces an unbound hyperon recoiling under the influence of a hyperon-nucleus potential. Thus the strength of the reaction is determined by a phase space calculation using Fermi-Dirac statistics and the hyperon-nucleus potential. It may be shown with a reasonable approximation [22] that the spectrum shape of the $K^- + A \rightarrow \pi^- + \Lambda + (A-1)$ reaction is parabolic having a maximum at $\bar{\omega}$ given by;

$$\bar{\omega} = M_\Lambda - M_N + (U_N - U_\Lambda) - (M_\Lambda - M_N)k_F^2/(4M_\Lambda M_N) + q^2/(2M_\Lambda) . \quad (2)$$

Here, M_Λ and M_N are the Λ and N masses, and U_Λ and U_N the well depths. The Fermi momentum is k_F and q and ω are the momentum, $q = p_K - p_\pi$,

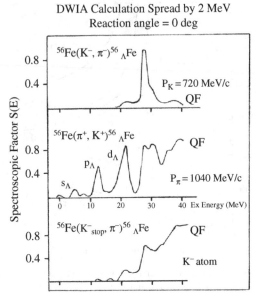

Fig. 8. Calculated spectroscopic factors for various reactions on an ^{56}Fe target illustrating the excitation strength to different states of the hypernucleus and QF production [5]. The energy scale in the figure is the excitation energy in the hypernucleus

and energy, $\omega = E_K - E_\pi$ transfers, respectively. Applying this analysis to the continuum data of several medium A hypernuclei, a Λ-nucleus well depth of $\approx 30\,\text{MeV}$ is extracted, Fig. 9.

As another example, Fig. 10 shows a fit to Σ-nucleus production data and is consistent with a strongly repulsive interaction. The best fit constrains the Σ to move in an complex, repulsive optical potential [23].

Hyperon interactions in the continuum spectrum should also include resonant behavior, i.e. nuclear structure information. Inclusion of nuclear structure can be treated by several methods [24, 25], the most common being the continuum shell model [26], where the QF and resonant behavior are simultaneously calculated.

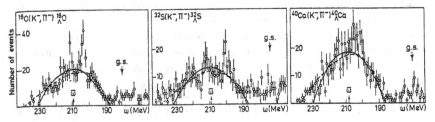

Fig. 9. The application of the Fermi gas model to the continuum structure of several medium A hypernuclei from which a Λ-nucleus well depth is extracted [22]

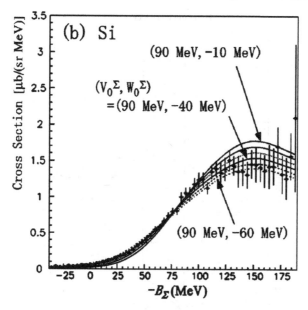

Fig. 10. The application of a complex optical potential to fit the Σ-nucleus final state interaction, showing strong repulsion and absorption [23]

3.4 Summary of the General Features of Production Mechanisms

The previous description of hypernuclear production assumes that the reaction occurs on a nucleon in a single particle state within a nucleus, leading to a hyperon that interacts with the spectator nucleons through a hyperon-nucleus potential well. If the momentum transfer is small the hyperon generally resides in the same angular momentum state (shell) as the original nucleon. For higher momentum transfers, there can be changes to the angular momentum structure of the resulting hypernucleus. In general, a hypernucleus is left in an excited or unbound state.

Note by observing Fig. 3, a hypernucleus is in its ground state when the hyperon and nucleons all reside in their lowest shell states, and this generally requires re-arrangements of the structure after production. The energy released in these rearrangements [27] can be removed by gamma rays or Auger neutron emission (see Fig. 11). In a hypernucleus, the neutron emission threshold can be lower than the Λ emission threshold, and in any event, nucleon emission can occur even from unbound Λ states. Thus the final hypernuclear species may differ from the one initially produced. Indeed the hypernuclear system may fission, producing a hypernucleus much lower in mass.

Therefore hypernuclei can be studied either by; 1) production mechanisms where the reaction is constrained by measuring the reaction products; or by

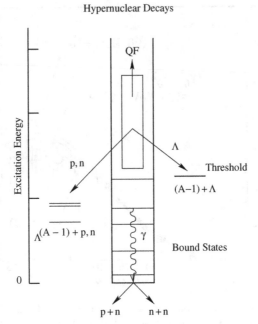

Fig. 11. A schematic representation of the decays of an excited hypernucleus, showing in particular the decay of highly excited states by Auger and gamma transitions

2) decay mechanisms where the production process may be ill (or not) determined, but measurement of the decay products is sufficient to identify a hypernucleus.

4 Beams and Detectors

4.1 Beams

Technology drives the experimental ability to collect information about any physics process, so the advent of intense beams of mesons, coupled with modern electronics changed the scope of hypernuclear investigations. Previously, experiments were limited to the observation of a few events due to low beam intensity and the tedious scanning of emulsion detectors. However, present high energy accelerators can produce protons and electrons at GeV energies where they can provide secondary beams of mesons and photons tailored for hypernuclear investigations. Figure 12 shows the intensity of various secondary particle beams, produced at zero degrees from a tungsten target [28, 29], as a function of the incident proton momentum. In this figure, the number of particles produced is normalized by the number of protons interacting in the target. This provides a more useful representation than

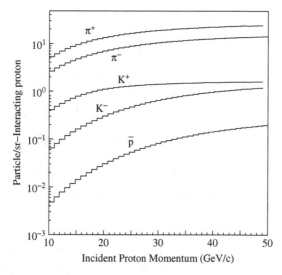

Fig. 12. Secondary particle production by energetic protons on a W target as a function of the incident proton momentum [28]

cross section, since much of the light meson production in particular, comes from heavier particle decays. In addition to mesons, the figure also shows the production intensity for anti-protons. Although hypernuclear production using anti-protons has been proposed [30], the potential use of such beams has not yet been established and will not be discussed here. The production and use of photon beams for hypernuclear research will be discussed in a later section.

Production targets for secondary beams are usually several interaction lengths thick, so that secondary particle production occurs from multiple interactions of the incident beam (and other particles produced by this beam) with the target. All secondary particles produced near zero degrees are collected by beamline transport magnets, and momentum selected before being focused onto a target where the hypernuclear reaction takes place. Figure 13 shows a typical, low-momentum kaon transport line that has parameters given in Table 3. The beamline has two $E \times B$ velocity separators which are tuned to pass kaons and remove pions, and a sextupole used to correct the transport optics [31, 32].

As is obvious from Fig. 12 pion production is significantly higher than kaon production, and experiments using kaons would be overwhelmed unless the pion flux is selectively attenuated by a π/K separator. In such a separator, electric fields as high as ± 225 kV are applied to electrostatic plates ≈ 10 cm apart, and the magnetic field is tuned to allow particles of a selected velocity to travel without deflection through the separation slits. The entire separator is enclosed in a pressurized tank containing SF_6 to limit sparking. A vertical slit, a few mm high, selects an enriched secondary beam of a particular mass.

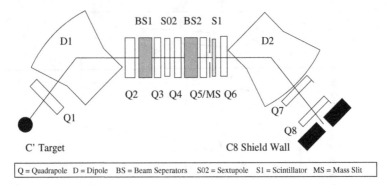

Fig. 13. A schematic drawing of a low-momentum kaon beam line showing π-K separation, BS1 and BS2 [32]

This "mass" slit then acts as the object for the magnetic transport into the experimental cave. The performance of a separator depends on the beam momentum, separation length, and optical quality of the beamline. In the case of the beamline described above, one finds a π^-/K^- ratio of $(5\text{–}10)/1$ for momenta near 800 MeV/c. Similar beamlines operated at higher momenta produce a π/K ratio of $1/1$.

Finally one should note that unstable particles have a mean free path length given by $\beta\gamma c\tau$. In the case of kaons, $c\tau$ is 3.71 m, and while this is boosted by the relativistic γ factor, the mean-life is not significantly changed for low-momentum kaons. On the other hand, the value of $c\tau$ for pions is 7.80 m and their decay is not so badly affected. Thus kaon beam lines, and/or the kaon detection systems must be made as short as possible.

4.2 Detectors

As described earlier, nuclear emulsion was the first detection system used to investigate hypernuclear events. The advantage of emulsion is its position

Table 3. The parameters of a typical kaon separated beamline [32]

Maximum Momentum - 820 MeV/c
Length - 19.6 m
Optics Corrected to 3rd Order
Maximum Primary Beam on Production Target - $30 \times 10^{12}\,\text{s}^{-1}$

Particle	Momentum (MeV/c)	Particles/s (a)	Purity
K$^+$	800	4.8×10^6	71%
K$^+$	Stopped	1.0×10^6	–

(a) Rate for 10×10^{12} particles/s, 25 GeV/c primary beam momentum, 9 cm Pt production target

and energy resolution, which allows detailed investigation of a reaction and its decay products. Indeed, emulsion is still the best detector in experiments designed to search for lower-momentum, exotic reactions where production and decay are not well known, or difficult to extract from background. Once a reaction is better understood, counter techniques can be designed which yield high statistics, better target selectivity, and extraction of contributing small reaction mechanisms. Counter detectors are required as beam intensities increase, and the improvement of electronic systems now allows much more detailed investigations to be undertaken.

In order to measure a hypernuclear spectrum one needs to determine the excitation energy (mass) of the recoiling hypernuclear system with sufficient resolution to separate states on the order of an MeV or so apart. Figure 14 shows that the Λ shells are separated by 5–10 MeV across the periodic table. However, states corresponding to excitations of the nuclear core with a Λ in a particular shell lie between these $\hbar\omega$ structures, and would be unresolved with a resolution level ≥ 1 MeV. In addition, hyperfine splitting of the $\hbar\omega$ single particle states would also be unresolved.

A typical spectrum for a (π^+, K^+) on a heavy target [7, 34] is shown in Fig. 15. The energy resolution of this spectrum is approximately 2.5 MeV and is due to contributions from the measurements of the momentum of the beam pion and reaction kaon by magnetic spectrometers, and energy straggling in the target. The Λ $\hbar\omega$ (neutron-hole, Λ-particle) shell structure is obvious (solid Gaussian curves). The dotted curves in the figure represent additional Gaussian strength that was included to fit the spectrum as shown by the heavier solid curve.

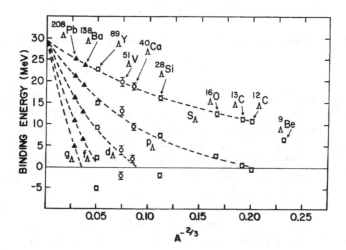

Fig. 14. The Energy levels of the major hypernuclear shells as a function of the $A^{-2/3}$ of the hypernucleus. The curves result from a calculation using an effective, density dependent potential for the Λ nucleus interaction [33]

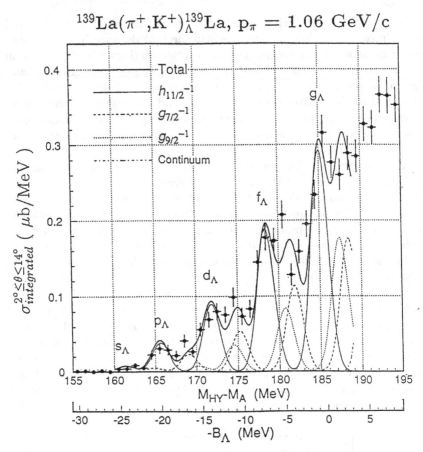

Fig. 15. The hypernuclear spectrum of $^{139}_{\Lambda}$La showing the major Λ shell structure [34]

This experiment used a typical magnetic spectrometer for the detection of reaction kaons, Fig. 16. One should note the particle identification system (PID) that is used to differentiate reaction pions and protons from kaons. This system consists of time-of-flight scintillators (SD), and threshold Cerenkov detectors. Without PID the kaon signal would be overwhelmed by background.

Another experiment has attempted to exploit the features of the $(K^{-}_{\text{stop}}, \pi^{-})$ reaction for hypernuclear production. A stopped K^{-} cascades through atomic orbitals by X-ray emission until it is captured by the nucleus [36]. The kaon capture process proceeds mainly with the emission of a pion forming the hypernucleus. Since the reaction occurs essentially at rest, the energy resolution of a hypernuclear state can be improved over measurements using a double spectrometer system, because in this case the momentum of the incident projectile does not have to be determined. However, energy straggling in the target seriously deteriorates the resolution, as low energy kaon beams have

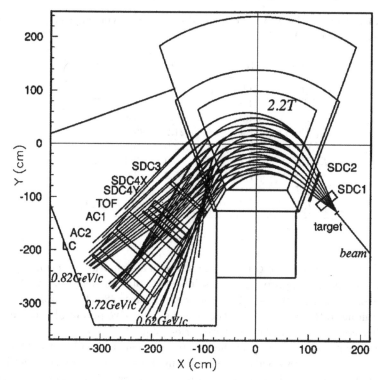

Fig. 16. A modern Kaon spectrometer showing typical particle trajectories and a particle identification system consisting of time-of-flight scintillators (SDXX) and Cerenkov detectors (ACX,LC) [35]

a range of momenta and must be brought to rest by energy loss in the target. This problem is addressed by the FINUDA detector which uses an almost mono-energetic K^- flux produced from the two body decay of ϕ mesons at rest, $\phi \rightarrow K^+K^-$. However, the momentum of the π^- released in the $(K^-_{\text{stop}}, \pi^-)$ reaction still must be determined by a magnetic spectrometer, limiting the energy resolution to $\geq 1\,\text{MeV}$.

The stopped reaction has higher momentum transfer than the in-flight reaction (see Fig. 7), and is much less selective, since kaon absorption generally leads to Σ rather than Λ production. This is shown in Table 4 which is taken from K^-_{stop} reactions in bubble chambers [37]. In this table the R factors are the branching fractions to a particular channel upon K^- capture. The ratio, R_n/R_p, is the ratio of captures on neutrons to protons, and the ratio, R_m, is the branching ratio for capture on multi-nucleons in the nucleus. Substantial background and various states are thus seen to be produced.

While energy resolution using magnetic spectrometers and meson beams is presently limited to no better than a few MeV, the energy of electromagnetic transitions between states can be measured to a few keV. The measurement

Table 4. Branching ratios $R(Y\pi)$ normalized to unity for hyperon production, using stopped K^- (in percent) [37]

Ratio	H	D	He	C	Ne
$R(\Lambda\pi^0)$	4.9	5.	6.2	4.4	3.4
$R(\Sigma^+\pi^-)$	14.9	30.	37.3	37.7	37.7
$R(\Sigma^-\pi^+)$	34.9	22.	10.9	16.8	20.4
$R(\Sigma^0\pi^0)$	21.4	23.	21.2	25.7	27.6
$R(\Lambda\pi^-)$	9.7	10.	12.6	8.7	6.7
$R(\Sigma^0\pi^-)$	7.1	5.	5.9	3.3	2.1
$R(\Sigma^-\pi^0)$	7.1	5.	5.9	3.3	2.1
R_n/R_p	0.31	0.25	0.32	0.18	0.12
R_m		0.01	0.16	0.19	0.23

of transition photons uses a coincidence between a formation reaction and the emission photon. It requires special solid-state, photon detectors with high photo-peak efficiency and rate handling capabilities [38]. The application of this technique to the extraction of the Λ p-shell effective potential is discussed in another contribution to this lecture series [39].

Experimentally, the energy of the emitted photon is spread by the Doppler boost due to the recoil velocity of the hypernucleus. This can be corrected using kinematics calculated from the energies and angles measured in the formation reaction, but there is a complication due to the continuous degradation of the velocity of the recoil as the stopping time in many cases is comparable to the electromagnetic transition lifetime. To correct for this spread, the stopping power as a function of time is used to correct the line shape of the signal. This not only improves the transition energy measurement, but provides the lifetime of a hypernuclear transition.

Finally, the coincidence technique can be applied to the detection of nucleon and meson emission due to the weak decay of the hypernucleus. Weak decay lifetimes are $\approx 0.2 \times 10^{-9}$ s, approximately the same as that of a free Λ, although a new non-mesonic weak-decay channel, $\Lambda + N \rightarrow N + N$, opens in nuclear matter. The lifetime for heavier hypernuclei is also found to be approximately independent of the hypernuclear mass. Initial $\Lambda N \rightarrow NN$ coincidence measurements between the formation and non-mesonic decay of a hypernucleus detected only one of the decay nucleons. While this information is useful, final state interactions between the decay nucleon and the residual nucleus distorted the spectrum and significantly affected the neutron to proton stimulated decay ratio, $\Gamma_n/\Gamma_p = \dfrac{\Lambda + n \rightarrow n + n}{\Lambda + p \rightarrow p + n}$. More recent measurements of both nucleons emitted in coincidence are in much better agreement with the theoretical predictions [40]. Figure 17 shows a detection system positioned around the target in a (π^+, K^+) reaction so that coincident particle emission from a hypernucleus weak decay can be detected.

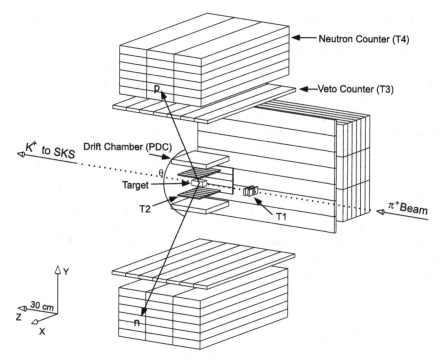

Fig. 17. The layout of a detection system to observe particle emission in the weak decay of a hypernucleus [40]

4.3 The (γ, K^+) Reaction

As an example of a specific production mechanism, we now discuss in more detail the (γ, K^+) reaction (see Fig. 6), as it is relatively new, and has the potential of achieving excellent energy resolution. Electro-production traditionally has been used for precision studies of nuclear structure, as the exchange of a photon can be accurately described by a first order perturbation calculation, and electron beams have excellent spatial and energy resolutions. Modern, continuous beam accelerators can now handle high singles-rate coincidence experiments, and although the cross section for kaon electro-production is some 2 orders of magnitude smaller than the (π^+, K^+) reaction, this can be compensated by increased beam intensity.

Generally the $(e, e'K^+)$ reaction has high spin-flip probability even at forward angles, and the momentum transfer is high (compare the curves in Fig. 7). Thus the resulting reaction is expected to predominantly excite spin-flip transitions to spin-stretched states of unnatural parity [41]. Such states are not strongly excited in mesonic production, and the electromagnetic process acts on a proton rather than a neutron creating proton-hole Λ-states, charge symmetric to those studied with meson beams. An additional advantage is that targets can be physically small and thin (10–$50\,\mathrm{mg/cm^2}$), allowing studies of almost any isotope.

In electro-production, the energy and momentum of the virtual photon are defined by $\omega = E_e - E_{e'}$ and $q = p_e - p_{e'}$, respectively. The experimental geometry is illustrated in Fig. 18. The four-momentum transfer of the electron to the virtual photon is then given by $Q^2 = \omega^2 - q^2$ ($Q^2 < 0$), which is chosen to be almost on the mass shell, i.e. a real photon. Since the elementary cross section, and particularly the nuclear form factor, fall rapidly with increasing $|Q^2|$, and the virtual photon flux is maximized for an electron scattering angle near zero degrees [42, 43], experiments must be done within a small angular range around the direction of the virtual photon.

The experimental geometry requires two spectrometers, one to detect the scattered electron which defines the virtual photon, and one to detect the kaon. Both of these spectrometers must be placed at extremely forward angles, Fig. 19. Because of this, a septum or splitting, magnet is needed to deflect the electron and kaon away from zero degrees into their respective spectrometers.

In the one photon approximation, the electro-production cross section can be expressed [44] by;

$$\frac{\partial^3 \sigma}{\partial E'_e \partial \Omega'_e \partial \Omega_K} = \Gamma \left[\frac{\partial \sigma_T}{\partial \Omega_K} + \epsilon \frac{\partial \sigma_L}{\partial \Omega_K} + \epsilon \frac{\partial \sigma_P}{\partial \Omega_K} \cos(2\phi) \right.$$
$$\left. + \cos(\phi)\sqrt{2\epsilon(1+\epsilon)} \frac{\partial \sigma_I}{\partial \Omega_K} \right].$$

Here ϕ is the out of plane angle, and the factor, Γ, is the virtual photon flux factor evaluated with electron kinematics in the lab frame. It has the form;

$$\Gamma = \frac{\alpha}{2\pi^2 |Q|^2} \frac{E_\gamma}{1 - \epsilon} \frac{E'_e}{E_e} .$$

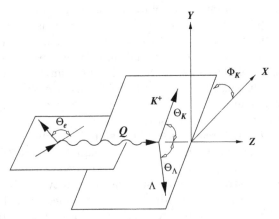

Fig. 18. The geometry of an (e, e', K^+) showing the incident electron beam, the virtual photon, and the reaction kaon

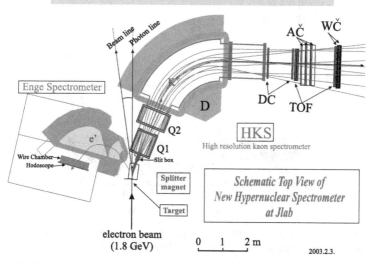

Fig. 19. A schematic layout of the apparatus to produce hypernuclei by the $(e, e'K^+)$ reaction [45]

In the above equation, ϵ is the polarization factor;

$$\epsilon = \left[1 + \frac{2q^2}{|Q|^2}\tan^2(\Theta_e/2)\right]^{-1}.$$

For virtual photons almost on the mass shell, $Q^2 \to 0$. The label on each of the cross section expressions represent transverse (T), longitudinal (L), polarization (P), and interference (I) terms. For real photons of course, $Q^2 \to 0$, so only the transverse cross section is non-vanishing. For the selected geometry, the electro-production cross section may also be replaced to good approximation, by the photo-production value.

Substantial numbers of pions, positrons, and protons are transmitted through the kaon spectrometer. Therefore excellent particle identification is required, not only in the hardware trigger, but also in the data analysis. The reconstructed missing mass of a hypernuclear state is a function of the beam energy, the momenta of the scattered electron and kaon, and the scattering angles. In a two-dimensional space defined by the electron and kaon momenta, the recoil missing mass is obtained by a projection of the events onto a locus line. Using an incorrect value of the beam energy or central momentum value for either spectrometer arm, results in an incorrect position and slope of the locus line, and therefore an incorrect kinematic position and width for various missing masses. An example of the binding energy spectrum for the $^{12}_{\Lambda}$B hypernucleus is shown in Fig. 20 along with the accidental background. The

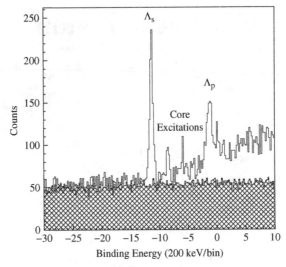

Fig. 20. The binding energy spectrum for $^{12}_{\Lambda}$B electro-produced from a ^{nat}C target. The solid histogram is the measured accidental background [46]

two prominent $\hbar\omega$, s- and p- shell, peaks are obvious in the spectrum. The energy resolution is $\approx 750\,\mathrm{keV}$ and can be improved in future experiments.

5 The Production of S$=-2$ Nuclear Systems

5.1 Background

Strangeness -2 could potentially be introduced directly into a nucleus by ranging a Ξ^- produced in a target, where the Ξ^- captures into an atomic orbital and interacts with the nuclear core. The direct production of a Ξ using the $(\mathrm{K}^-, \mathrm{K}^+)$ reaction requires a two step process, with the transfer of a strange quark from the K^- and the associated production of an $s\bar{s}$ pair. In general this reaction has high probability for production of QF Ξ, and although one could hope to use QF Ξ to induce S $= -2$ reactions on other nuclei, most decay before they range or interact. Even those that do interact, for example by $\Xi + \mathrm{N} \to \Lambda + \Lambda$ conversion, provide about $30\,\mathrm{MeV}$ of energy that is equally shared between the two Λs in most cases, leading to the escape of one or both from the nucleus. However, this is not the complete story as will be discussed in the next subsections.

Rather general theoretical calculations predict that multi-strange hadronic matter (SHM), having a strangeness to baryon fraction ≈ -1 and a charge to baryon fraction ≈ 0, might be at least meta-stable at densities 2–3 times the nuclear density. Such SHM decays by the weak interaction back to a system of nucleons [49].

It has also been proposed that strange quark matter (SQM) having an approximately equal number of u, d, and s quarks might be stable even with respect to the strong interaction [47, 48]. Thus this has motivated an intensive search for an S = −2 dibaryon composed of two u, d, and s quarks, which presumably can take advantage of $SU(3)_f$ symmetry to produce a deeply bound system, the H particle [50]. The H in the hadronic limit might also be considered as a bound (or low lying resonant) composition of $\Lambda\Lambda$, $\Sigma\Sigma$, and ΞN hyperons. However, there has been no evidence for an H particle either below or above the $\Lambda\,\Lambda$ threshold in experiments looking at both production and decay channels [51, 52]. Neutral, long lived particles are not easily observed by measuring their trajectories or decay products. However a missing mass experiment would remain sensitive to the production of an H that might be either stable or have a resonant structure.

At higher temperatures, the Λ-N interaction is relevant to the cooling of the particle "plasma" formed in relativistic heavy ion collisions [5, 41, 68, 69]. The coalescence model has had modest success in predicting production cross sections for mesons, hyperons, and a limited number of light single Λ hypernuclei. The model allows the residual baryons in the central rapidity region to cool thermally, coalescing those that have sufficiently low momentum and spatial separation within phase space. It has also been applied to the production of double-Λ systems including the H particle [5], but one has no way at present to validate these calculations.

5.2 The Double Λ Systems Formed in (K^-, K^+) Reaction

On the other hand, double Λ hypernuclei have conclusively been observed through the sequential pion decays [53, 54, 55, 56, 58] of their s-shell Λs, but there is little experimental information on such systems. Only four $\Lambda\Lambda$ hypernuclei are reported, as single events in five experiments. One of the observations could not uniquely identify the hypernucleus, Table 5, and the $_{\Lambda\Lambda}^{6}$He hypernucleus was seen in two different experiments. Emulsion detectors have played a crucial role in the discovery of these events.

The extracted binding energies from these events are not internally consistent. However, the consistency of the data is improved by neglecting the

Table 5. Observed $\Lambda\Lambda$ Hypernuclei

Hypernucleus	Detection	Reference
$_{\Lambda\Lambda}^{10}$Be	Emulsion	[53]
$_{\Lambda\Lambda}^{6}$He	Emulsion	[54]
$_{\Lambda\Lambda}^{10}$Be; or $_{\Lambda\Lambda}^{13}$B	Emulsion	[55]
$_{\Lambda\Lambda}^{6}$He	Hybrid Emulsion	[56]
$_{\Lambda\Lambda}^{4}$H	Counter	[58]

earlier, and retaining the more recent, $_{\Lambda\Lambda}^{6}$He event. Theoretically the NSC97 Nijmegen model, [57], then reproduces the experimental binding energy of this reduced data set, although other Nijmegen models cover a range of possibilities. If the reduced data set is accepted, the $\Delta B_{\Lambda\Lambda}$ is changed from approximately 4.6 to 1 MeV; i.e from a strong binding to a weak one. The quantity, $\Delta B_{\Lambda\Lambda}$ is defined by the equations;

$$\text{Mass}(_{\Lambda\Lambda}A) = \text{Mass}(A - 2) + 2\,\text{Mass}(\Lambda) - B_{\Lambda\Lambda}, \tag{3}$$
$$\Delta B_{\Lambda\Lambda} = B_{\Lambda\Lambda} - 2\,B_{\Lambda}(A - 1).$$

Thus $\Delta B_{\Lambda\Lambda}$ represents the additional binding energy in the double Λ system, which comes from the mutual interaction of the two Λs in the nucleus either directly or indirectly by altering the nuclear core. The weak $\Delta B_{\Lambda\Lambda}$ value confirms the fact that an H, at least as a hadronic state, should not exist. In fact it is questionable if even $_{\Lambda\Lambda}^{4}$H is bound [59, 60, 61]. However if this hypernucleus is bound, the measure of its binding energy would be an important constraint on the Λ-Λ interaction.

The NSC97 model is weakly attractive in the Λ-Λ and N-Ξ channel, but strongly attractive in the Ξ-Ξ, Σ-Σ and Σ-Ξ channels. To the extent that this model represents the strange hyperon-hyperon interactions, one would expect strong coupling between mixtures of hadronic states. For example, a first order phase transition between N$\Lambda\Xi$ and N$\Sigma\,\Xi$ hadronic matter is predicted at about 3 times nuclear density as the strangeness fraction $f_s = -S/A$ is increased, Fig. 21. Much more detailed discussion is available in the references concerning the Λ-Λ and Λ-Ξ interaction and the value of $\Delta B_{\Lambda\Lambda}$.

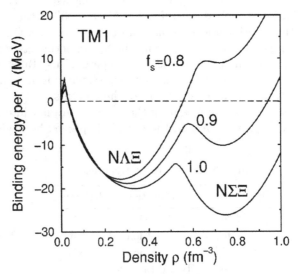

Fig. 21. A calculation showing a transition between N$\Lambda\Xi$ and N$\Sigma\,\Xi$ matter as a function of matter density for various strangeness fractions [62]

Clearly more examples of double Λ hypernuclei are needed, and await an intense facility producing K^- at GeV energies, such as J-PARC now under construction in Japan.

5.3 Nuclear Systems Containing a Ξ^-

The existence of Ξ hypernuclei is not established, although there are several emulsion events [63, 64] which can be interpreted as $\Xi^- + C \rightarrow {}^4_\Lambda H + {}^9_\Lambda Be$. In one of these events [63], the binding of an intermediate Ξ state was interpreted as $B_\Xi = 0.54$ MeV, probably indicating that fission occurred from an atomic level. In the other event [64], the binding energy was indeterminate because the recoiling hypernuclei could have been in an excited state. However, a binding energy as large as 3.7 MeV was possible.

The missing mass spectrum in another experiment using the ${}^{nat}C(K^-, K^+)$ reaction, [65, 66] was observed with very poor resolution. Individual levels could not be ascertained, but from QF analysis, the Ξ-nucleus well depth was estimated to be ≈ 15 MeV. If this is correct, a bound state should exist in the $\Xi - {}^{11}B$ system with a binding energy of about 6 MeV. The width of the Ξ states depends on the strong conversion $\Xi N \rightarrow \Lambda\Lambda$. However this decay channel can be Pauli blocked if Λs fill the decay shells [49], providing Ξ hypernuclear states sufficiently narrow for spectroscopic measurements.

It was pointed out in [59] that due to the possible strong $\Lambda - \Xi$ attraction proposed by the NSC97 model, the $S = -3$ hypernuclei ${}^6_{\Lambda\Xi}H$ and ${}^6_{\Lambda\Xi}He$ may provide the onset of Ξ stability in nuclear mater.

6 Systems with Multiple Strangeness

As the matter density increases to that in neutron stars, hyperons, and perhaps their dissociation into quarks, would become absolutely stable. As discussed above, it is expected that roughly equal compositions of u, d, and s quarks, leading to a strangeness fraction $f_s = -S/A \approx 1$ and charge fraction $f_Q = Z/A \approx 0$ occur in hadronic systems at these densities and significantly affect the radius and maximum mass of such stars. The particle composition of neutron star matter has been calculated by several authors [47, 48, 67, 70, 71, 72, 73]. Figure 22 shows a typical example of the particle composition of a neutron star as a function of its density. It is, of course, speculated that the composition changes from the interior to exterior of the star, that is from a core of quark matter to hyperon-nucleon matter to conventional neutron star matter.

When the remnant of a supernova collapses it undergoes a conversion process that results in a neutron star, a hyperon-nucleon star, and a quark star (or more likely, a combination of all three as a function of the radial density). At each stage of collapse, energy is released as the phase transition occurs, and there is the potential of a secondary explosion if this phase transition is first

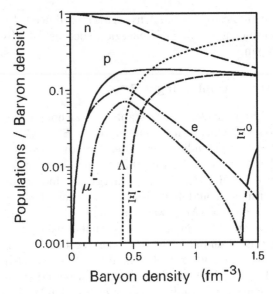

Fig. 22. A typical calculation of the particle composition of a neutron star assuming in this case a strongly repulsive Σ-nucleus interaction [71]

order [74]. In any event the radius of a rotating neutron star is dependent on its mass, which may provide a way to validate these calculations. The present maximum mass of a hyperon-neutron star is predicted to be less than approximately 1.8 solar masses [75], and observational astronomy may provide limits to verify this prediction.

However, input to the model requires knowledge of the unknown hyperon-baryon potentials. Hypernuclear physics provides the possibility of at least extracting the Λ-N, and Λ-Λ interaction at normal nuclear densities, and these can serve as a normalization point, in order to extrapolate the SU(3)$_f$ interaction to the matter-densities found in neutron stars.

7 Concluding Remarks

In summary, the investigation of strangeness in nuclear systems is not merely an extension of conventional nuclear physics. Certainly one cannot, nor would one want to, reproduce the wealth of information that has been accumulated on conventional nuclei. Indeed, the strangeness degree of freedom allows the nuclear particles to rearrange by taking advantage of SU(3)$_f$ flavor symmetry, in order to maximize the nuclear binding energy [76]. Thus a hypernuclear system can better illuminate features which are more obscured in conventional nuclear systems. Such questions as isobar mixing, charge symmetry breaking, and quark confinement are more important, and thus more evident, in strange hadronic matter. Also of relevance are various other features of hypernuclei such as weak decay, and the stability of multi-strange hadronic systems. With respect to this latter issue,the conclusive observation of doubly strange nuclear systems and the extraction of Λ-Λ and Ξ-N interactions are vital. What is now needed is a series of precision studies with high resolution where level positions and weak decay dynamics can be compared to theory.

Acknowledgments

The author acknowledges the instruction and guidance of many colleagues over the years who have contributed to my knowledge of nuclear physics. I also acknowledge the partial support of the US Department of Energy under grant DE-FG02-94ER40836. Hypernuclear physics has been an exciting field of exploration for 50 years, and with its re-birth in Europe and Japan, should have a promising future.

References

1. A. K. Wroblewski: Acta Phys. Pol. B **35**, 901 (2004)
2. M. Danysz and J. Pniewski: Phil. Mag. **44**, 348 (1953)
3. M. Gell-Mann: Phys. Rev. **92**, 833 (1953); A. Pais: Phys. Rev. **86**, 663 (1952)
4. D. H. Davis and J. Pniewski: Contemp. Phys. **27**, 91 (1986); D. H. Davis, Nucl. Phys. A **754**, 3c (2005)
5. H. Bando, T. Motoba, and J. Žofka: Int. J. Mod. Phys. A **5**, 4021 (1990)
6. R. H. Chrien: *Table of observed Λ hypernuclei* (National Data Center, 2000)
7. O. Hashimoto and H. Tamura, Prog. Part. Nucl. Phys. **57**, 564 (2006)
8. B. F. Gibson and E. V. Hungerford: Phys. Rep. **257**, 350 (1995)
9. M. M. Nagels, Th. A. Rijken, and J. J. de Swart: Phys. Rev. D **15**, 2547 (1977)
10. M. M. Nagels, Th. A. Rijken, and J. J. de Swart: Phys. Rev. D **20**, 1633 (1979)
11. P. M. M. Maessen, Th. A. Rijken, and J. J. de Swart: Phys. Rev. C **40**, 2226 (1989)
12. Th. A. Rijken, V. G. J. Stoks, and Y. Yamamoto: Phys. Rev. C **59**, 21 (1999)
13. A. Reuber, K. Holinde, and J. Speth: Nucl. Phys. A **570**, 543 (1994)
14. C. B. Dover, D. J. Millener, and A. Gal: Phys. Rep. **184**, 1 (1989)
15. A. Gal: Adv. Nucl. Phys. **8**, 1 (1975)
16. A. Gal, J. M. Soper, and R. H. Dalitz: Ann. Phys. (NY) **63**, 53 (1971); Ann. Phys. (NY) **72**, 445 (1972); Ann. Phys. (NY) **113**, 79 (1978)
17. R. H. Dalitz and A. Gal: Ann. Phys. (NY) **116**, 167 (1978)
18. J. V. Noble: Phys. Lett. B **89**, 325 (1980)
19. B. F. Gibson and D. R. Lehman: Phys. Rev. C **22**, 2024 (1980); Phys. Rev. C **37**, 679 (1988)
20. A. Nogga, H. Kamada, and W. Glockle: Phys. Rev. Lett. **88**, 172501 (2002)
21. J. Hufner, S. Y. Lee, and H. A. Weidenmuller: Nucl. Phys. A **234**, 429 (1974)
22. R. H. Dalitz and A. Gal: Phys. Lett. B **64**, 154 (1976)
23. P. K. Saha et al: Phys. Rev. C **70**, 044613 (2004)
24. R. E. Chrien, E. Hungerford, and T. Kishimoto: Phys. Rev. C **35**, 1589 (1987)

25. T. Motoba et al: Phys. Rev. C **38**, 1322 (1988)
26. D. Halderson and R. J. Philpott: Phys. Rev. C **37**, 1104 (1988)
27. A. Likar, M. Rosina, and B. Povh: Z. Phys. A **324**, 35 (1986)
28. J. R. Sanford and C. L. Wang: BNL Report 11479 (1967)
29. D. Berley: BNL Report 50579 (1976)
30. M. Agnello, F. Ferro, and F. Iazzi: arXiv:hep-ex/0405061
31. D. M. Lazarus: BNL Report 50579 (1976)
32. P. Pile: private communication, BNL Proposal E949 (2004)
33. D. J. Millener, C. B. Dover, and A. Gal: Phys. Rev. C **38**, 2700 (1988)
34. T. Hasegawa et al: Phys. Rev. C **53**, 1210 (1996)
35. T. Fukuda et al: Nucl. Instrum. Methods Phys. Res. A **361**, 485 (1995)
36. A. Zenoni: Proc. International School of Physics (Enrico Fermi) CLVIII (2005) p. 183
37. C. Vander Velde-Wilquet et al: Nuovo Cim. A **39**, 538 (1977)
38. H. Tamura et al: Phys. Rev. Lett. **84**, 5963 (2000); **86**, 1982 (2001)
39. D. J. Millener: in this volume
40. O. Outa: Proc. International School of Physics (Enrico Fermi) CLVIII (2005) p. 219; T. Nagae: in this volume
41. T. Motoba, M. Sotona, and K. Itonaga: Prog. Theor. Phys. Suppl. **117**, 123 (1994)
42. C. E. Hyde-Wright, W. Bertozzi, and J. M. Finn: Proc. CEBAF Summer Workshop, Newport News, VA (1985)
43. G. H. Xu and E. V. Hungerford: Nucl. Instrum. Methods Phys. Res. A **501**, 602 (2003)
44. H. Thom: Phys. Rev. **151**, 1322 (1966)
45. O. Hashimoto, spokesperson for E01-011 at the Jefferson National Laboratory
46. V. Rodrigues: Spectroscopic study of Λ-hypernuclei beyond the p-shell region: The HKS experiment at J-Lab, PhD Thesis, University of Houston, Houston, Texas (2006)
47. A. R. Bodmer: Phys. Rev. D **4**, 1601 (1971)
48. E. Witten: Phys. Rev. D **30**, 272 (1984)
49. J. Schaffner et al: Phys. Rev. Lett. **71**, 1328 (1993); J. Schaffner et al: Ann. Phys. (NY) **235**, 35 (1994)
50. R. L. Jaffe: Phys. Rev. Lett. **38**, 195 (1977)
51. L. Lee et al: Nucl. Phys. A **684**, 598c (2001)
52. S. Aoki et al: Phys. Rev. Lett. **65**, 1729 (1990)
53. M. Danysz et al: Nucl. Phys. **49**, 121 (1963)
54. D. J. Prowse: Phys. Rev. Lett. **17**, 782 (1966)
55. S. Aoki et al: Prog. Theor. Phys. **85**, 87 (1991)
56. H. Takahashi et al: Phys. Rev. Lett. **87**, 212502 (2001)
57. V. G. J. Stoks and Th. A. Rijken: Phys. Rev. C **59**, 3009 (1999)
58. J. K. Ahn et al: Phys. Rev. Lett. **87**, 132504 (2001)
59. I. N. Filikhin and A. Gal: Phys. Rev. Lett. **89**, 172502 (2002); Phys. Rev. C **65**, 041001(R) (2002)
60. H. Nemura, Y. Akaishi, and K. S. Myint: Phys. Rev. C **67**, 051001(R) (2003)
61. A. Gal: Nucl. Phys. A **754**, 91c (2005)
62. J. Schaffner-Bielich and A. Gal: Phys. Rev. C **62**, 034311 (2000)
63. S. Aoki et al: Prog. Theor. Phys. **89**, 493 (1993)
64. S. Aoki et al: Phys. Lett. B **355**, 1306 (1998)

65. T. Fukuda et al: Phys. Rev. C **58**, 1306 (1998)
66. P. Khaustov et al: Phys. Rev. C **61**, 054603 (2000)
67. A. Gal and C. B. Dover: Nucl. Phys. A **585**, 1c (1995)
68. M. A. C. Lamont: arXiv:nucl-ex/0608017
69. M. Sano and M. Wakai: Prog. Theor. Phys. Suppl. **117**, 99 (1994)
70. N. K. Glendenning and J. Schaffner-Bielich: Phys. Rev. C **58**, 1298 (1998)
71. N. K. Glendenning: Phys. Rev. C **64**, 025801 (2001)
72. I. Bednarek and P. Manka: J. Phys. G **31**, 1009 (2005)
73. N. K. Glendenning, F. Weber, and S. A. Moszkowski: Phys. Rev. C **45**, 844 (1992)
74. A. Bhattacharyya et al: arXiv:astro-ph/0606523
75. F. Weber: contribution to the IXth International Conference on Hypernuclear and Strange Particle Physics, Mainz, Germany (2006), to be published
76. E. H. Auerbach et al: Phys. Rev. Lett. **47**, 1110 (1981)

Hypernuclear Gamma-Ray Spectroscopy and the Structure of p-shell Nuclei and Hypernuclei

D. J. Millener

Brookhaven National Laboratory, Upton, NY 11973, USA
millener@bnl.gov

Abstract. Information on $^7_\Lambda$Li, $^9_\Lambda$Be, $^{10}_\Lambda$B, $^{11}_\Lambda$B, $^{12}_\Lambda$C, $^{15}_\Lambda$N, and $^{16}_\Lambda$O from the Ge detector array Hyperball is interpreted in terms of shell-model calculations that include both Λ and Σ configurations with p-shell cores. It is shown that the data puts strong constraints on the spin dependence of the ΛN effective interaction.

1 Introduction

The structure of Λ hypernuclei – i.e. many-body systems consisting of neutrons, protons, and Λ particles – is an interesting subject in its own right. However, the finer details of the structure of single-Λ hypernuclei, particularly the energy spacings of doublets formed by the coupling of a Λ in the lowest s orbit to a nuclear core state with non-zero spin, provide information on the spin dependence of the effective ΛN interaction. This is important because data on the free YN interaction are very sparse and essentially limited to spin-averaged s-wave scattering.

The spectroscopy of Λ hypernuclei has been reviewed recently by Hashimoto and Tamura [1]. The 'workhorse' reactions used to produce Λ hypernuclei have been the (K^-, π^-) (strangeness exchange) and (π^+, K^+) (associated production) reactions that convert a neutron into a Λ. The elementary $n(K^-, \pi^-)\Lambda$ and $n(\pi^+, K^+)\Lambda$ reactions have predominantly non-spin-flip character at the incident beam energies used.

The first information on Λ hypernuclei came from their production via K^- mesons stopped in emulsion followed by their π^--mesonic weak decay [2]. These studies provided Λ separation energies (B_Λ values) up to $A \sim 15$. These could be accounted for by a Λ-nucleus potential of Woods-Saxon shape with a depth of about 30 MeV. A number of ground-state spins were determined from angular correlation studies and weak-decay branching ratios, γ rays from excited states of $^4_\Lambda$H and $^4_\Lambda$He were seen, and so was proton emission from excited states of $^{12}_\Lambda$C. Currently, counter experiments with stopped K^- mesons are being performed at Frascati [3].

D. J. Millener: *Hypernuclear Gamma-Ray Spectroscopy and the Structure of p-shell Nuclei and Hypernuclei*, Lect. Notes Phys. **724**, 31–79 (2007)
DOI 10.1007/978-3-540-72039-3_2   © Springer-Verlag Berlin Heidelberg 2007

The momentum transfer to the hypernucleus is rather small in the forward direction in (K^-, π^-) reactions near the beam momenta of $\sim 800\,\text{MeV/c}$ used at CERN and BNL and the cross sections for $\Delta L = 0$ transitions are large. Because the cross sections are proportional to the spectroscopic factors for neutron removal from the target, the cross sections are largest when a high-spin neutron orbit is just filled and the produced Λ occupies the same orbit. Such transitions have been observed in selected nuclei up to ^{209}Bi (see [1]).

The (π^+, K^+) reaction has been used at $p_\pi = 1.05\,\text{GeV/c}$ where the elementary cross section peaks strongly. The momentum transfer is high $(q \sim 350\,\text{MeV/c})$ and the reaction selectively populates high-spin states. The cross sections are smaller than for the (K^-, π^-) reaction but the count rates for producing Λ hypernuclei can be comparable because more intense pion beams can be used. Transitions from nodeless high-spin neutron orbits can populate the full range of nodeless bound, and just unbound, Λ orbitals and have been used to map out the spectrum of Λ single-particle energies for selected nuclei up to ^{208}Pb (see [1]). This information has provided a rather precise characterization of the Λ-nucleus potential. The best resolution, obtained at KEK using the SKS spectrometer and a thin ^{12}C target, is $1.45\,\text{MeV}$ [4].

The free Λ decays into a nucleon and a pion with a lifetime of $263\,\text{ps}$. In a hypernucleus, the low-energy nucleon produced via this decay mode is Pauli blocked and the process $\Lambda N \rightarrow NN$ rapidly dominates with increasing mass number. Nevertheless, the measured weak decay lifetimes of Λ hypernuclei remain around $200\,\text{ps}$. This means that particle-bound excited states of Λ hypernuclei normally decay electromagnetically. Then it is possible to make use of the excellent resolution of γ-ray detectors to measure the spacings of hypernuclear levels. The earliest measurements were made with NaI detectors but the superior (few keV) resolution of Ge detectors has been exploited since 1998 in the form of the Hyperball, a large-acceptance Ge detector array. A series of experiments on p-shell targets has been carried out at KEK and BNL using the $(\pi^+, K^+\gamma)$ and $(K^-, \pi^-\gamma)$ reactions, respectively [1]. As well as γ-ray transitions between bound states of the primary hypernucleus, γ-ray transitions are often seen from daughter hypernuclei formed by particle emission (most often a proton) from unbound states of the primary hypernucleus.

The results of these γ-ray experiments are interpreted in terms of nuclear structure calculations with a parametrized effective YN interaction as input.

2 The ΛN (YN) Effective Interaction

The hyperon-nucleon interaction involves the coupled ΛN and ΣN channels, as illustrated in Fig. 1. The diagrams in Fig. 1 make the point that the direct ΛN–ΛN interaction does not contain a one-pion exchange contribution because of isospin conservation (except for electromagnetic violations via Λ–Σ^0 mixing) while the coupling between the ΛN and ΣN channels does. For this reason,

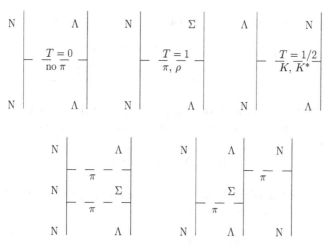

Fig. 1. Diagrams showing the important features of the coupled $\Lambda N - \Sigma N$ strangeness -1 interaction for isospin $1/2$. The last diagram shows the two-pion exchange three-body interaction

the ΛN interaction is relatively weak and there is reason to believe that the three-body interaction in a hypernucleus could be relatively important.

The free-space interactions are obtained as extensions of meson-exchange models for the NN interaction by invoking, e.g., a broken flavor SU(3) symmetry. The most widely used model is the Nijmegen soft-core, one-boson-exchange potential model known as NSC97 [5]. The six versions of this model, labelled NSC97a..f, cover a wide range of possibilities for the strength of the central spin-spin interaction ranging from a triplet interaction that is stronger than the singlet interaction to the opposite situation. An extended soft-core version (ESC04) has recently been published [6]. Effective interactions for use in a nuclear medium are then derived through a G-matrix procedure [5, 6].

The ΛN effective interaction can be written (neglecting a quadratic spin-orbit component) in the form

$$V_{\Lambda N}(r) = V_0(r) + V_\sigma(r)\, s_N \cdot s_\Lambda + V_\Lambda(r)\, l_{N\Lambda} \cdot s_\Lambda + V_N(r)\, l_{N\Lambda} \cdot s_N + V_T(r)\, S_{12}\,, \quad (1)$$

where V_0 is the spin-averaged central interaction, V_σ is the difference between the triplet and singlet central interactions, V_Λ and V_N are the sum and difference of the strengths of the symmetric spin-orbit (LS) interaction $l_{N\Lambda} \cdot (s_\Lambda + s_N)$ and antisymmetric spin-orbit (ALS) interaction $l_{N\Lambda} \cdot (s_\Lambda - s_N)$, and V_T is the tensor interaction with (C^2 is a normalized spherical harmonic)

$$S_{12} = 3(\sigma_N \cdot \hat{r})(\sigma_\Lambda \cdot \hat{r}) - \sigma_N \cdot \sigma_\Lambda$$
$$= \sqrt{6}\, C^2(\hat{r}) \cdot [\sigma_N, \sigma_\Lambda]^2 \,. \quad (2)$$

For the Λ in an s orbit, $l_{N\Lambda}$ is proportional to l_N [7]. The effective ΛN–ΣN and ΣN–ΣN interactions can be written in the same way.

Each term of the potential in (1) can be written in the form

$$V_k(r)\, \mathcal{L}^k \cdot \mathcal{S}^k = V_k(r)\, (-)^k \widehat{k}[\mathcal{L}^k, \mathcal{S}^k]^0 \,, \tag{3}$$

where k is the spherical tensor rank of the orbital and spin operators and $\widehat{k}^2 = 2k + 1$.

So-called YNG interactions, in which each term of the effective interaction is represented by an expansion in terms of a limited number of Gaussians with different ranges,

$$V(r) = \sum_i v_i\, e^{-r^2/\beta_i^2} \,, \tag{4}$$

are often used for the central and spin-orbit components with the following form used for the tensor component,

$$V_T(r) = \sum_i v_i\, r^2\, e^{-r^2/\beta_i^2} \,. \tag{5}$$

For example, the G-matrix elements from a nuclear matter calculation have been parametrized in this way, in which case the YNG interactions have a density dependence (through k_F).

Given the interaction in YNG, or some other, radial form, two-body matrix elements that define the interaction for a shell-model calculation can be calculated using a chosen set of single-particle radial wave functions. In the following, the procedure is sketched for harmonic oscillator radial wave functions in the case of equal mass particles. There are techniques to calculate the two-body matrix elements for any (e.g., Woods-Saxon) radial wave functions but the harmonic oscillator case illustrates where the important contributions come from and suggests ways in which the interaction can be parametrized in terms of the radial matrix elements themselves.

Separating the space and spin variables in (1) and (3) using (64)

$$\langle l_1 l_2 L S | V | l_1' l_2' L' S' \rangle^{JT} = \sum_k (-)^{L'+S+J} \begin{Bmatrix} L & L' & k \\ S' & S & J \end{Bmatrix}$$
$$\times \widehat{L} \langle l_1 l_2 L || V_k(r)\mathcal{L}^k || l_1' l_2' L' \rangle \widehat{S} \langle S || \mathcal{S}^k || S' \rangle \,. \tag{6}$$

Harmonic oscillator wave functions have the unique property that a transformation exists from the individual particle coordinates r_1, r_2 to the relative and center of mass coordinates $(r_1 - r_2)/\sqrt{2}$, $(r_1 + r_2)/\sqrt{2}$ of the pair [8]. This transformation and another application of (64) result in an expression in terms of radial integrals in the relative coordinate $r = |r_1 - r_2|$

$$\langle l_1 l_2 L || V_k(r)\mathcal{L}^k || l_1' l_2' L' \rangle = \sum_{N_c L_c l l'} (-)^{l+L_c+k+L'} \widehat{L}\widehat{l} \begin{Bmatrix} L_c & l' & L' \\ k & L & l \end{Bmatrix}$$
$$\times \langle nl N_c L_c, L | n_1 l_1 n_2 l_2, L \rangle \langle n'l' N_c L_c, L' | n_1' l_1' n_2' l_2', L' \rangle$$
$$\times \langle nl | V_k(r) | n'l' \rangle \langle l || \mathcal{L}^k || l' \rangle \,, \tag{7}$$

where the number of quanta associated with coordinate r is given by $q = 2n+l$ ($n = 0, 1, ..$) and energy conservation $q_1 + q_2 = q + Q_c$ fixes n (similarly n'). The reduced matrix elements (see Appendix A) of the orbital and spin operators are listed in Table 1. The radial integral can in turn be expressed in terms of Talmi integrals I_p

$$\langle nl|V(r)|n'l'\rangle = \sum_p B(nl, n'l'; p) I_p . \qquad (8)$$

The harmonic oscillator radial relative wave function is a polynomial in r' times $\exp(-r'^2)$ where $r' = |\mathbf{r_1} - \mathbf{r_2}|/\sqrt{2}\,b$ and $b^2 = \hbar/m\omega$. Then,

$$I_p = \frac{2}{\Gamma(p + 3/2)} \int_0^\infty r^{2p} e^{-r^2} V(\sqrt{2}rb)\, r^2\, dr . \qquad (9)$$

For a Gaussian potential, $V(r) = V_0 \exp(-r^2/\mu^2)$, with $\theta = b/\mu$,

$$I_p = \frac{V_0}{(1 + 2\,\theta^2)^{p+3/2}} . \qquad (10)$$

For the case of a Λ in an s orbit attached to a light nucleus, the expressions for the matrix elements of each component of the interaction are shown in Table 2. In this simple case, the I_p are equal to the relative matrix elements in the angular momentum states denoted by p (the superfluous superscripts denote the interaction in even or odd relative states). For the nuclear p shell, there are just five $p_N s_\Lambda$ two-body matrix elements formed from $p_{1/2} s_{1/2}(0^-, 1^-)$ and $p_{3/2} s_{1/2}(1^-, 2^-)$ (alternatively, $L = 1$ and $S = 0, 1$). This means that the five radial integrals $\overline{V}$, Δ, S_Λ, S_N, and T associated with each operator in Table 2 can be used to parametrize the ΛN effective interaction. In Appendix B, the $p_N s_\Lambda$ two-body matrix elements are given in terms of the parameters in both LS and jj coupling.

A comprehensive program for the shell-model analysis of Λ binding energies for p-shell hypernuclei was set out by Gal, Soper, and Dalitz [7], who also included the three-body double-one-pion-exchange ΛNN interaction shown in Fig. 1. This interaction does not depend on the spin of the Λ and was characterized by a further five radial integrals. Dalitz and Gal went on to consider the formation of p-shell hypernuclear states via (K^-, π^-) and (K^-, π^0) reactions and the prospects for γ-ray spectroscopy based on the decay of these

Table 1. Two-particle reduced matrix elements of orbital and spin operators

	1	$s_N \cdot s_\Lambda$	LS	ALS	Tensor
$\widehat{S}\langle S\|\mathcal{S}^k\|S'\rangle$	$\widehat{S}$	$\widehat{S}[2S(S+1) - 3]/4$	$\delta_{SS'}\widehat{S}\sqrt{2}$	$(-)^S(1 - \delta_{SS'})\sqrt{3}$	$\widehat{S}\sqrt{20/3}$
$\langle l\|\mathcal{L}^k\|l'\rangle$	1	1	$\delta_{ll'}\sqrt{l(l+1)}$	$\delta_{ll'}\sqrt{l(l+1)}$	$\sqrt{6}\langle l020\|l'0\rangle$

Table 2. ΛN (YN) parameters

$V_{N\Lambda}$	$s_N s_\Lambda$	$p_N s_\Lambda$	$^7_\Lambda$Li values (MeV)
V_0	I_0^e	$\overline{V} = \frac{1}{2}(I_0^e + I_1^o)$	(-1.22)
$V_\sigma s_N \cdot s_\Lambda$	I_0^e	$\Delta = \frac{1}{2}(I_0^e + I_1^o)$	0.480
$V_\Lambda l_N \cdot s_\Lambda$		$S_\Lambda = \frac{1}{2}I_1^o$	-0.015
$V_N l_N \cdot s_N$		$S_N = \frac{1}{2}I_1^o$	-0.400
$V_T S_{12}$		$T = \frac{1}{3}I_1^o$	0.030

states [9]. Unfortunately, knowledge of the ground-state B_Λ values plus a few constraints from known ground-state spins was insufficient to provide definitive information on the spin-dependence of the ΛN interaction.

The most direct information on the spin dependence of the ΛN effective interaction comes from the spacing of s_Λ doublets based on core states with non-zero spin. These spacings depend on the parameters Δ, S_Λ, and T that are associated with operators that involve the Λ spin. The energy separations of states based on different core states depend on S_N, but these separations can also be affected by the three-body interaction. Millener, Gal, Dover, and Dalitz [10] made estimates for Δ, S_Λ, S_N, and T using new information, particularly on γ-ray transitions in $^7_\Lambda$Li and $^9_\Lambda$Be [11], together with theoretical input from YN interaction models. These estimates were close to the values given in the right-hand column of Table 2 (the bracketed value for $\overline{V}$ is not fitted) which fit the now-known energies of the four bound excited states of $^7_\Lambda$Li (see Sect. 4). An alternative set of parameters was proposed by Fetisov, Majling, Žofka, and Eramzhyan [12] who were motivated by the non-observation of a γ-ray transition from the ground-state doublet of $^{10}_\Lambda$B in the first experiment using Ge detectors [13].

Experiments with the Hyperball, starting in 1998 with ^{7}Li$\,(\pi^+, K^+\gamma)\,$ $^7_\Lambda$Li at KEK and ^{9}Be$\,(K^-, \pi^-\gamma)\,$ $^9_\Lambda$Be at BNL, have provided the energies of numerous γ-ray transitions, together with information on relative intensities and lifetimes. The progress of the theoretical interpretation in terms of shell-model calculations has been summarized at HYP2000 [14] and HYP2003 [15]. By the latter meeting, Σ degrees of freedom were being included explicitly through the inclusion of both Λ and Σ configurations in the shell-model basis.

The most convincing evidence for the importance of Λ–Σ coupling comes from the s-shell hypernuclei and this is described in the following section. This is followed by a discussion of $^7_\Lambda$Li in Sect. 4. Because the LS structure of the p-shell core nuclei plays an important role in picking out particular combinations of the spin-dependent ΛN parameters, Sect. 5 is devoted to a general survey of p-shell calculations, spectra, and wave functions. This information is used in subsequent sections that are devoted to the remaining hypernuclei, up to $^{16}_\Lambda$O, for which data, particularly from γ rays, exists.

3 The s-shell Λ Hypernuclei

The data on the s-shell hypernuclei is shown in Table 3. The B_Λ values come from emulsion data [2] and the excitation energies from γ rays observed following the stopping of negative kaons in ^{6}Li and ^{7}Li [16].

The spin-spin component of the ΛN interaction contributes to the splitting of the 1^+ and 0^+ states of $^4_\Lambda$H and $^4_\Lambda$He. In the case of simple $s^3 s_\Lambda$ configurations, the contribution is given by the radial integral of the spin-spin interaction in s states (the Δ in Table 2) using

$$\sum_i s_i \cdot s_\Lambda = S_c \cdot s_\Lambda$$

$$= \frac{1}{2}[S^2 - S_c^2 - s_\Lambda^2] . \tag{11}$$

However, it has long been recognized as a problem to describe simultaneously the binding energies of the s-shell hypernuclei with a central ΛN interaction and that this problem might be solved by Λ–Σ coupling which strongly affects the 0^+ states of the A = 4 hypernuclei. Recently, there has been a clear demonstration of these effects and it was found that the spin-spin and Λ–Σ coupling components of the NSC97e and NSC97f interactions give comparable contributions to the 1^+–0^+ doublet splitting [17]. Subsequent studies using a variational method with Jacobi-coordinate Gaussian-basis functions [18], Faddeev-Yakubovsky calculations [19], and stochastic variational calculations with correlated Gaussians [20] have confirmed and illustrated various aspects of the problem.

Akaishi et al [17] calculated G-matrices for a small model space of s orbits only, writing two-component wave functions for either the 0^+ or the 1^+ states of $^4_\Lambda$He (or $^4_\Lambda$H) with isospin T = 1/2

$$|^4_\Lambda\text{He}\rangle = \alpha s^3 s_\Lambda + \beta s^3 s_\Sigma . \tag{12}$$

The Σ component is 2/3 Σ^+ and 1/3 Σ^0 for $^4_\Lambda$He (2/3 Σ^- and 1/3 Σ^0 for $^4_\Lambda$H). The off-diagonal matrix element

$$v(J) = \langle s^3 s_\Lambda, J|V|s^3 s_\Sigma, J\rangle \tag{13}$$

Table 3. Data on the s-shell Λ hypernuclei

Hypernucleus	$J^\pi(gs)$	B_Λ (MeV)	J^π	E_x (MeV)
$^3_\Lambda$H	$1/2^+$	0.13(5)		
$^4_\Lambda$H	0^+	2.04(4)	1^+	1.04(5)
$^4_\Lambda$He	0^+	2.39(3)	1^+	1.15(4)
$^5_\Lambda$He	$1/2^+$	3.12(2)		

can be derived from the $\Lambda N - \Sigma N$ G matrix for $0s$ orbits, where V is used for the potential representing the G matrix interaction, by splitting one nucleon off from the s^3 configurations using the fractional parentage expansion

$$|s^3\rangle = \sum_{S(T)} \frac{1}{\sqrt{2}} (-)^{1+S} |[s^2(TS), s](1/2\,1/2)\rangle , \qquad (14)$$

where $TS = 0\,1$ or $1\,0$. Coefficients of fractional parentage (cfp) specify how to construct a fully antisymmetric n-particle state from antisymmetric $(n-1)$-particle states coupled to the nth particle [cf. (41)]. In this simple case, the magnitude of the cfp is determined by the symmetry with respect to T and S and the phase appears twice and cancels out in the problem at hand. Then, by recoupling on either side of (13) (see Appendix A),

$$v(J) = \frac{3}{2} \sum_{S\bar{S}} U(S1/2\,J1/2, 1/2\,\bar{S})^2 U(T1/2\,1/2\,0, 1/2\,1/2)$$
$$\times\ U(T1/2\,1/2\,1, 1/2\,1/2)\langle ss_\Lambda, \bar{S}|V|ss_\Sigma, \bar{S}\rangle , \qquad (15)$$

where the factor of 3 appears because any one of the three s-shell nucleons can be chosen from an antisymmetric wave function. Specializing to the case of $J = 0$

$$v(0) = \frac{3}{2}\,{}^3V - \frac{1}{2}\,{}^1V$$
$$= \overline{V} + \frac{3}{4}\Delta , \qquad (16)$$

where

$$\overline{V} = \frac{1}{4}\,{}^1V + \frac{3}{4}\,{}^3V , \qquad \Delta = {}^3V - {}^1V \qquad (17)$$

Similarly,

$$v(1) = \frac{1}{2}\,{}^3V + \frac{1}{2}\,{}^1V$$
$$= \overline{V} - \frac{1}{4}\Delta . \qquad (18)$$

Taking round numbers derived using the 20-range Gaussian interaction of [17] which represents NSC97f yields ${}^3V = 4.8\,\text{MeV}$ and ${}^1V = -1.0\,\text{MeV}$, which give $\overline{V} = 3.35\,\text{MeV}$ and $\Delta = 5.8\,\text{MeV}$. Then, $v(0) = 7.7\,\text{MeV}$ and $v(1) = 1.9\,\text{MeV}$. In a simple 2×2 problem, the energy shifts of the Λ-hypernuclear states are given by $\sim v^2(J)/\Delta E$ with $\Delta E \sim 80\,\text{MeV}$ (and the admixture $\beta \sim -v(J)/\Delta E$). Thus, the energy shift for the 0+ state is $\sim 0.74\,\text{MeV}$ while the shift for the 1^+ state is small. The result is close to that for the NSC97f interaction in Fig. 1 of [17].

The same method can be used to obtain the singlet and triplet contributions of the Λ interaction for all the s-shell hypernuclei in the case of simple s-shell nuclear cores. The results are given in Table 4. The expressions in terms of $\overline{V}$ and Δ can be written down by inspection.

Table 4. Contributions of singlet and triplet interactions to s-shell hypernuclei

Hypernucleus	$J^\pi(gs)$	^3V and ^1V	$\overline{V}$ and Δ
$^3_\Lambda$H	$1/2^+$	$3/2\,^1V + 1/2\,^3V$	$2\overline{V} - \Delta$
$^3_\Lambda$H	$3/2^+$	$2\,^3V$	$2\overline{V} + 1/2\Delta$
$^4_\Lambda$He, $^4_\Lambda$H	0^+	$3/2\,^1V + 3/2\,^3V$	$3\overline{V} - 3/4\Delta$
$^4_\Lambda$He, $^4_\Lambda$H	1^+	$1/2\,^1V + 5/2\,^3V$	$3\overline{V} + 1/4\Delta$
$^5_\Lambda$He	$1/2^+$	$^1V + 3\,^3V$	$4\overline{V}$

4 The $^7_\Lambda$Li Hypernucleus

The first p-shell hypernucleus with particle-stable excited states that can be studied by γ-ray spectroscopy is $^7_\Lambda$Li and it is of interest to compare the effects of the ΛN spin-spin interaction and Λ–Σ coupling in $^7_\Lambda$Li with those in $^4_\Lambda$H and $^4_\Lambda$He.

The low-lying states of $^7_\Lambda$Li consist of a Λ in an s orbit coupled (weakly) to a ^{6}Li core. Only the $1^+; 0$ $(J^\pi; T)$ ground state of ^{6}Li is stable with respect to deuteron emission but the Λ brings in extra binding energy and the lowest particle-decay threshold for $^7_\Lambda$Li is $^5_\Lambda$He+d at 3.94(4) MeV derived from

$$S_d(^7_\Lambda\text{Li}) = S_d(^6\text{Li}) + B_\Lambda(^7_\Lambda\text{Li}) - B_\Lambda(^5_\Lambda\text{He})$$
$$= 1.475 + 5.58(3) - 3.12(2) \,, \tag{19}$$

where the B_Λ values (errors in parentheses) come from emulsion studies [2].

Figure 2 shows the spectrum of $^7_\Lambda$Li determined from experiments KEK E419 [21, 22] and BNL E930 [23] with the Hyperball detector. The four γ-rays seen in [21] – all except the $7/2^+ \rightarrow 5/2^+$ transition – show that the state based on the $0^+; 1$ state of ^{6}Li is bound at an excitation energy of 3.88 MeV. Only the $5/2^+ \rightarrow 1/2^+$ transition was previously known from an experiment at BNL with NaI detectors [11]. Because the $3/2^+$ state is expected to be weakly populated in the (π^+, K^+) reaction, much of the intensity of the 692-keV γ-ray transition comes from feeding via the γ-ray transition from the $1/2^+; 1$ level. The $7/2^+ \rightarrow 5/2^+$ doublet transition was seen in γ γ coincidence with the $5/2^+ \rightarrow 1/2^+$ transition following ^{3}He emission from highly-excited states of $^{10}_\Lambda$B (the s-hole region) produced via the (K^-, π^-) reaction on ^{10}B [23].

Shell-model calculations for p-shell hypernuclei start with the Hamiltonian (Y can be a Λ or a Σ)

$$H = H_N + H_Y + V_{NY} \,, \tag{20}$$

where H_N is some empirical Hamiltonian for the p-shell core, the single-particle H_Y supplies the ~ 80 MeV mass difference between Λ and Σ, and V_{NY} is the YN interaction. The two-body matrix elements of the YN interaction between states of the form $(p_N s_Y)$ can be parametrized in the form given in Table 2 (see Appendix B). This form applies to the direct ΛN interaction, the ΛN–ΣN coupling interaction, and the direct ΣN interaction for both

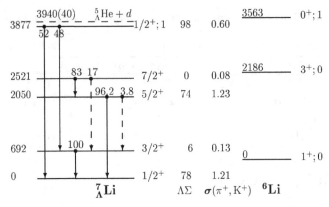

Fig. 2. The spectrum of $^7_\Lambda$Li determined from experiments KEK E419 and BNL E930 with the Hyperball detector. All energies are in keV. The solid arrows denote observed γ-ray transitions. The γ-ray branching ratios are theoretical and the dashed arrows correspond to unobserved transitions. For each state of $^7_\Lambda$Li, the calculated energy shifts due to Λ–Σ coupling and the calculated relative populations via the (π^+, K^+) reaction are given [24]. The core states of ^{6}Li are shown on the right

isospin $1/2$ and $3/2$ (which is included in the calculations). The shell-model basis states are chosen to be of the form

$$|(p^n \alpha_c J_c T_c, j_Y Y)JT\rangle , \qquad (21)$$

where the hyperon is coupled in angular momentum and isospin to eigenstates of the p-shell Hamiltonian for the core. This is known as a weak-coupling basis and, indeed, the mixing of basis states in the hypernuclear eigenstates is generally very small. In this basis, the core energies are taken from experiment where possible and from the p-shell calculation otherwise.

For $^7_\Lambda$Li, the basis states are of the form $|p^2 s_\Lambda\rangle$ and $|p^2 s_\Sigma\rangle$. The p^2 wave functions for ^{6}Li are close to the LS-coupling limit, as can be seen from Table 5 where wave functions are given for all three of Cohen and Kurath's interactions [25] and two other interactions fitted to p-shell data. As discussed in more detail in Sect. 5, the central interaction is attractive in spatially even (S and D) states and repulsive in (P) odd states. The $3^+;0$ state in Fig. 2 is the lowest member of an $L=2$, $S=1$ (^{3}D) triplet completed by a $2^+;0$ state at 4.31 MeV and a $1^+;0$ state at 5.65 MeV; the $2^+;1$ (^{1}D) state is at 5.67 MeV. The $L=1$ admixtures are largely through the one-body spin-orbit interaction. The $p_{1/2} - p_{3/2}$ splittings at $A=5$ are small (0.14–1.29 MeV) for the Cohen and Kurath interactions (the p states are unbound at $A=5$ and the $p_{1/2}$ energy is poorly defined). The fit69 interaction has a larger splitting of 3.5 MeV while the fit5 interaction, the one used in the hypernuclear calculation, has an intermediate value of 1.8 MeV.

The structure of the core nucleus means that the $3/2^+$ and $7/2^+$ states are mainly $L=0$, $S=3/2$ and purely $L=2$, $S=3/2$, respectively. This accounts for

Table 5. Wave functions for $A = 6$ using a number of different interactions. The $2^+; 0$ and $3^+; 0$ states are uniquely ^{3}D

J_n^π	$^{(2S+1)}$L	fit69	fit5	CK616	CK816	CKPOT
$1_1^+; 0$						
	^{3}S	0.9873	0.9906	0.9576	0.9484	0.9847
	^{3}D	−0.0422	−0.0437	−0.2777	−0.3093	−0.1600
	^{1}P	−0.1532	−0.1298	−0.0761	−0.0703	−0.0685
$1_2^+; 0$						
	^{3}S	0.0287	−0.0347	−0.2810	−0.3082	−0.1426
	^{3}D	−0.9007	−0.9987	−0.9591	−0.9510	−0.9673
	^{1}P	0.4334	0.0708	−0.0354	0.0259	0.2096
$0_1^+; 1$						
	^{1}S	0.9560	0.9909	0.9997	0.9999	0.9946
	^{3}P	0.2935	0.1348	0.0247	−0.0137	0.1036
$2_1^+; 1$						
	^{1}D	0.8760	0.9827	0.9486	0.9959	0.9839
	^{3}P	0.4824	0.3148	0.1854	0.0905	0.1789

their low population in the (π^+, K^+) reaction which is dominantly non-spin-flip (the ^{7}Li ground state has L = 1, S = 1/2, J = 3/2). The $1/2^+$ states are mainly L = 0, S = 1/2 while the $5/2^+$ state is 7/9 S = 1/2 and 2/9 S = 3/2 in the LS limit for the core. In this limit, it is easy to derive the contribution of each of the spin-dependent ΛN parameters to the binding energies.

These contributions for the full shell-model calculation are given in Table 6 as the coefficients of each of the ΛN effective interaction parameters. In the LS limit for the ground-state doublet, only Δ contributes while for the excited-state doublet all terms contribute. For the $1/2^+; 1$ state, there would be no contributions. However in the realistic case, S_N contributes substantially for the predominantly L = 0 cases. This is because the associated operator $l_N \cdot s_N$ connects the L = 0 and L = 1 basis states of the core giving rise to a linear dependence on the amplitude of the L = 1 admixture. This admixture is quite

Table 6. Contributions of the spin-dependent ΛN terms to the binding energies of the five bound states of $^7_\Lambda$Li given as the coefficients of each of the ΛN effective interaction parameters. In the $\Lambda\Sigma$ column the gains in binding energy due to Λ–Σ coupling are given in keV (same as in Fig. 2)

$J_i^\pi; T$	$\Lambda\Sigma$	Δ	S_Λ	S_N	T
$1/2^+; 0$	78	−0.975	−0.025	0.242	0.080
$3/2^+; 0$	6	0.486	0.013	0.253	−0.205
$5/2^+; 0$	74	−0.796	−1.165	0.980	1.177
$7/2^+; 0$	0	0.500	1.000	1.000	−1.200
$1/2^+; 1$	98	−0.002	0.002	0.453	−0.005

sensitive to the model for the p-shell core (see Table 5). The hypernuclear shell-model states are very close to the weak-coupling limit. For the $5/2^+$ state, there is a 1.28% admixture based on the the $2^+;0$ core state (because the 2^+ and 3^+ core states share the same L and S). Otherwise, the intensity of the dominant basis state is $> 99.7\%$.

Table 7 re-expresses the same information in terms of energy differences between states and gives the actual energy contributions for the parameter set

$$\Delta = 0.430 \quad S_\Lambda = -0.015 \quad S_N = -0.390 \quad T = 0.030 . \tag{22}$$

This parameter set is chosen to reproduce the $^7_\Lambda$Li spectrum which it does quite well, as can be seen by comparing the energies in the last column of Table 7 with the experimental energies at the left of Fig. 2. Note that an increase in one or both of the 'small' parameters S_Λ and T could reduce the excited-state doublet splitting to the experimental value of 471 keV. Also that these two parameters have to take small values if they have the same signs as in (22). Tighter constraints on these parameters come from the spectra of heavier p-shell hypernuclei (see later).

Returning to the LS limit, the coefficient of Δ for the ground-state doublet, derived from (11), is $3/2$. A similar evaluation using the LS structure of the members of the excited-state doublet gives $7/6$ for the coefficient of Δ. In this case, the full expression for the splitting of the excited-state doublet is [9]

$$\Delta E = \frac{7}{6}\Delta + \frac{7}{3}S_\Lambda - \frac{14}{5}T . \tag{23}$$

This expression can be derived in a variety of ways using the results in Appendix A or Appendix B but perhaps most easily by multiplying the coefficients of Δ, S_Λ and T for the $7/2^+$ state in Table 6, for which twice the 2^- two-body matrix element [(66) or (67)] enters because the angular momenta are stretched for the $7/2^+$ state, by $7/3$ because their contribution measures the shift from the centroid $2\overline{V} + S_N$ of the $7/2^+$ and $5/2^+$ levels.

Table 7. Energy spacings in $^7_\Lambda$Li. ΔE_C is the contribution of the core level spacing. The first line in each case gives the coefficients of each of the ΛN effective interaction parameters as they enter into the spacing while the second line gives the actual energy contributions to the spacing in keV

$J_i^\pi - J_f^\pi$	ΔE_C	$\Lambda\Sigma$	Δ	S_Λ	S_N	T	ΔE
$3/2^+ - 1/2^+$			1.461	0.038	0.011	−0.285	
	0	72	628	−1	−4	−9	693
$5/2^+ - 1/2^+$			0.179	−1.140	0.738	1.097	
	2186	4	77	17	−288	33	2047
$1/2^+ - 1/2^+$			0.972	−0.026	0.211	−0.085	
	3565	−20	418	0	−82	−3	3886
$7/2^+ - 5/2^+$			1.294	2.166	0.020	−2.380	
	0	74	557	−32	−8	−71	494

The $\langle p_N s_\Lambda | V | p_N s_\Sigma \rangle$ matrix elements were calculated from the YNG interaction SC97f(S) of [17] using harmonic oscillator wave functions with $b = 1.7\,\text{fm}$. These matrix elements were multiplied by 0.9 to simulate the Λ-Σ coupling of SC97e(S) and thus the observed doublet splitting for ${}^4_\Lambda\text{He}$ (see [17]). In the same parametrization as for the ΛN interaction

$$\overline{V}' = 1.45 \quad \Delta' = 3.04 \quad S'_\Lambda = -0.085 \quad S'_N = -0.085 \quad T' = 0.157 \,. \qquad (24)$$

The YNG interaction has non-central components but the dominant feature is a strong central interaction in the ${}^3S$ channel reflecting the second-order effect of the strong tensor interaction in the ΛN–ΣN coupling. Because the relative wave function for a nucleon in a p orbit and a hyperon in an s orbit is roughly half s state and half p state, the matrix elements coupling Λ-hypernuclear and Σ-hypernuclear configurations are roughly a factor of two smaller than those for the A = 4 system. Because the energy shifts for the Λ-hypernuclear states are given by $v^2/\Delta E$, where v is the coupling matrix element and $\Delta E \sim 80$ MeV, the shifts in p-shell hypernuclei will be roughly a quarter of those for A = 4 in favorable cases; e.g. 150 keV if the Λ-Σ coupling accounts for about half of the A = 4 splitting. For T = 0 hypernuclei, the effect will be smaller because the requirement of a T = 1 nuclear core for the Σ-hypernuclear configurations brings in some recoupling factors which are less than unity. For example, in the case of the ${}^7_\Lambda\text{Li}$ ground state

$$\langle p^2(L=0\,S=1\,T=0)\,s_\Lambda, J = \tfrac{1}{2} \,|\, V \,|\, p^2(L=0\,S=0\,T=1)\,s_\Sigma, J = \tfrac{1}{2}\rangle$$

$$= 2\sum_{\bar{S}} U(1/2\,1/2\,1/2\,1/2, 1\bar{S})U(1/2\,1/2\,1/2\,1/2, 0\bar{S})\langle ps_\Lambda, \bar{S}|V|ps_\Sigma, \bar{S}\rangle$$

$$= \frac{\sqrt{3}}{2}\,({}^3V - {}^1V) = \frac{\sqrt{3}}{2}\,\Delta' \,. \qquad (25)$$

Putting in the value for Δ' from (24) and taking the actual value for ΔE of $\sim 88.5\,\text{MeV}$ gives 78 keV for the energy shift (cf. Table 6). The result depends only on Δ' because $\overline{V}'$ connects only states with the same core and because the spin-spin term is required to connect the core states in (25). In fact, the coefficients of the Λ-Σ coupling parameters depend on isovector one-body density-matrix elements connecting the core states (essentially $\langle \sigma\tau \rangle$ for Δ').

A comparison of the ground-state doublet splitting for ${}^7_\Lambda\text{Li}$ with that for ${}^4_\Lambda\text{He}$ (and ${}^4_\Lambda\text{H}$) using modern YN interactions with the same Monte Carlo [26, 27], or other few-body, methods for both should provide a tight constraint on the strength of the Λ-Σ coupling because the contributions to the doublet spacings are very different in the two cases.

5 The p-shell Nuclei

The structure of the spin-dependent operators in (1) means that their effects are most easily seen in an LS coupling basis. In particular, either L = 0 or

S=0 for the core isolates one of the parameters (Δ or S_Λ, respectively). The example of $^7_\Lambda$Li illustrated this point and is also a case in which calculations can be made by hand. In the general case, a shell-model calculation for the p shell is made with a phenomenological interaction fitted to p-shell data. The wave functions are expanded on a basis set (maximum dimension 14)

$$H\Psi = E\Psi \qquad \Psi = \sum_i a_i \Phi_i , \qquad (26)$$

where Φ could be expressed in jj coupling

$$\Phi = |p^m_{3/2}(J_1 T_1)p^{n-m}_{1/2}(J_2 T_2); JT\rangle \qquad (27)$$

or LS coupling (the Wigner supermultiplet scheme)

$$\Phi = |p^n[f]KLSJT\rangle . \qquad (28)$$

The purpose of the present section is to illustrate the structure of p-shell nuclei in terms of the latter basis. This basis turns out to be very good in the sense that the wave functions for low-energy states are frequently dominated by one basis state, or just a few basis states. This aids in the physical interpretation of the structure. From the hypernuclear point of view, the contributions of Δ and S_Λ depend only on the intensities of L and S in the total wave function and these can be obtained in the weak-coupling limit from a knowledge of the core wave function in an LS basis.

H is defined by two single-particle energies and 15 two-body matrix elements. In terms of the relative coordinates of a pair of nucleons, s, p and d states are possible for two p-shell nucleons. There are 6 central matrix elements (one in each relative state for S = 0, 1) which are attractive in spatially even states and repulsive in odd states. The central part of the Millener–Kurath interaction [28] (a single-range Yukawa interaction with $b/\mu = 1.18$, potential strengths in MeV) illustrates this point (the superscripts are $2T + 1$ and $2S + 1$)

$$V^{11} = 32.0 \qquad V^{31} = -26.88 \qquad V^{13} = -44.8 \qquad V^{33} = 12.8 . \qquad (29)$$

There are 6 vector matrix elements, 2 arising from spin-orbit interactions in triplet p and d states and 4 from antisymmetric spin-orbit (ALS) interactions that connect two-body states with S = 0 and S = 1 (these are not present in the free interaction for identical baryons). Finally, there are 3 tensor matrix elements in triplet p and d states and connecting triplet s and d, states.

The above approach is exemplified by the classic Cohen and Kurath [25] fits to p-shell energy levels in terms of a constant set of single-particle energies and two-body matrix elements. The assumption of an A-independent interaction is a reasonable one for the p shell because the rms charge radii of p-shell nuclei are rather constant, as shown in Table 8. This is basically because the p-shell

Table 8. Root-mean-square charge radii (fm) of stable p-shell nuclei

^{6}Li	^{7}Li	^{9}Be	^{10}B	^{11}B	^{12}C	^{13}C	^{14}C	^{14}N	^{15}N
2.57	2.41	2.52	2.45	2.42	2.47	2.44	2.56	2.52	2.59

nucleons become more bound as more particles are added to the shell and the rms radii of the individual orbits tend to stabilize as nucleons are added.

Cohen and Kurath obtained three different interactions by fitting binding energies relative to ^{4}He after the removal of an estimate for the Coulomb energy. These interactions were designated as (8–16)2BME, (6–16)2BME, and (8–16)POT where the mass ranges fitted are specified and POT means that the 4 ALS matrix elements out of the 17 parameters were set to zero. In the course of the hypernuclear studies described here (and for other reasons) many fits have been made to a modern data base of p-shell energy-level data. Only well determined linear combinations of parameters (considerably less than 17) defined by diagonalizing $\partial\chi^2/\partial x_i\partial x_j$ have been varied where χ^2 measures the deviation of the theoretical and experimental energies in the usual way and the x_i are the parameters. The fit69 and fit5 interactions in Table 5 are examples fitted to data on the A = 6–9 nuclei with only the central and one-body interactions varied for fit69 and with the tensor interaction and the one-body spin-orbit splitting fixed for fit5.

In the supermultiplet basis, $[f]$KL label representations of SU(3) $\supset$ R(3) in the orbital space and $[\widetilde{f}]\beta$TS label representations of SU(4) $\supset$ SU2 $\times$ SU2 in the spin-isospin space. Actually, $[f] = [f_1 f_2 f_3]$ labels representations of U(3) with $f_1 \geq f_2 \geq f_3$ and $f_1 + f_2 + f_3 = n$ and can be represented pictorially by a Young diagram with f_1 boxes in the first row, f_2 in the second and f_3 in the third. For a totally antisymmetric wave function, $[\widetilde{f}]$ must be the conjugate of $[f]$ and is obtained by interchanging the rows and columns of the Young diagram. The length of the rows is then restricted to four. In an oscillator basis, there is one quantum per particle in the p-shell and $[f]$ labels also the symmetries of the quanta and the wave functions have an SU(3) symmetry labelled by $(\lambda\mu) = (f_1 - f_2\ f_2 - f_3)$ with K and L given by

$$K = \mu, \mu - 2, ..., 1 \text{ or } 0$$
$$L = K, K + 1, ..., K + \lambda$$
$$L = \lambda, \lambda - 2, ..., 1 \text{ or } 0 \text{ if } K = 0 . \tag{30}$$

For convenience, the allowed quantum numbers for p^n configurations are given in Table 9. For $12 - n$ particles, the L and TS quantum numbers are the same and $(\lambda\mu) \rightarrow (\mu\lambda)$.

In the following subsections, the spectra of selected p-shell nuclei for $6 \leq A \leq 14$ are presented and discussed in relation to the supermultiplet structure of their p-shell wave functions (see the tabulations covering the energy levels

Table 9. Quantum numbers for p-shell nuclei

U(3)	SU(3)	L	U(4)	(TS)
[2]	(20)	0,2	[11]	$(01)(10)$
[11]	(01)	1	[2]	$(00)(11)$
[3]	(30)	1,3	[111]	$(\frac{1}{2}\frac{1}{2})$
[21]	(11)	1,2	[21]	$(\frac{1}{2}\frac{1}{2})(\frac{1}{2}\frac{3}{2})(\frac{3}{2}\frac{1}{2})$
[111]	(00)	0	[3]	$(\frac{1}{2}\frac{1}{2})(\frac{3}{2}\frac{3}{2})$
[4]	(40)	0,2,4	[1111]	(00)
[31]	(21)	1,2,3	[211]	$(01)(10)(11)$
[22]	(02)	0,2	[22]	$(00)(11)(02)(20)$
[211]	(10)	1	[31]	$(01)(10)(11)(12)(21)$
[41]	(31)	1,2,3,4	[1]	$(\frac{1}{2}\frac{1}{2})$
[32]	(12)	1,2,3	[221]	$(\frac{1}{2}\frac{1}{2})(\frac{1}{2}\frac{3}{2})(\frac{3}{2}\frac{1}{2})$
[311]	(20)	0,2	[311]	$(\frac{1}{2}\frac{1}{2})(\frac{1}{2}\frac{3}{2})(\frac{3}{2}\frac{1}{2})(\frac{3}{2}\frac{3}{2})$
[221]	(01)	1	[32]	$(\frac{1}{2}\frac{1}{2})(\frac{1}{2}\frac{3}{2})(\frac{3}{2}\frac{1}{2})(\frac{3}{2}\frac{3}{2})(\frac{1}{2}\frac{5}{2})(\frac{5}{2}\frac{1}{2})$
[42]	(22)	0,2²,3,4	[11]	$(01)(10)$
[411]	(30)	1,3	[2]	$(00)(11)$
[33]	(03)	1,3	[222]	$(00)(11)$
[321]	(11)	1,2	[321]	$(01)(10)(11)^2(02)(20)(12)(21)$
[222]	(00)	0	[33]	$(01)(10)(12)(21)(03)(30)$

of light nuclei [29] for more experimental information). Many of these nuclei form the nuclear cores of hypernuclei discussed in detail in later sections.

5.1 The Central Interaction

The central interaction gives the bulk of the binding energy in p-shell nuclei. It turns out to be essentially diagonal in the supermultiplet basis and can be represented by 5 SU(4) invariants.

$$H = 1.56\,n - 1.79\sum_{i<j} I_{ij} - 3.91\sum_{i<j} P_{ij} + 0.59\,\boldsymbol{L}^2 - 1.08\,\boldsymbol{S}^2 + 0.59\,\boldsymbol{T}^2 \,. \quad (31)$$

Here, the term linear in n includes the centroid energy of the $p_{3/2}$ and $p_{1/2}$ orbits at A $=5$ and takes care of the (constant) one-body terms that arise from $\boldsymbol{L}^2$, $\boldsymbol{S}^2$, and $\boldsymbol{T}^2$. The two-body identity operator counts the number of pairs $n(n-1)/2$ and the space-exchange operator counts the difference between the numbers of spatially symmetric and antisymmetric pairs $n_{\rm s} - n_{\rm a}$ given by

$$\langle\,[f]\,|\sum_{i<j} P_{ij}\,|\,[f]\,\rangle = \frac{1}{2}\sum_i f_i(f_i - 2i + 1)\,. \quad (32)$$

A rule of thumb is that this can be read off the Young diagram by summing the number of pairs for each row and subtracting the number of pairs for each column. The relationship of the space-exchange operator to the quadratic Casimir operator for SU(4) is also worth noting,

$$\sum_{i<j} P_{ij} = 2n - \frac{1}{8}n^2 - \frac{1}{2}C(\mathrm{SU}(4)),\qquad(33)$$

where

$$C(\mathrm{SU}(4)) = \frac{1}{2}\sum_{i<j}\boldsymbol{\sigma}_i\cdot\boldsymbol{\sigma}_j\,\boldsymbol{\tau}_i\cdot\boldsymbol{\tau}_j + \boldsymbol{S}^2 + \boldsymbol{T}^2 + \frac{9}{4}n.\qquad(34)$$

Since there are only 6 independent central matrix elements, there are only 6 independent operators to represent them. The remaining one would connect the space and spin-isospin spaces, e.g. $\boldsymbol{l}_i\cdot\boldsymbol{l}_j\,\boldsymbol{\tau}_i\cdot\boldsymbol{\tau}_j$. All the operators in (31) are SU(3) scalars except for $\boldsymbol{L}^2$ which transforms as a mixture of $(0\,0)$ and $(2\,2)$ tensors. An SU(3) tensor expansion of the central interaction contains four scalars and two $(2\,2)$ tensors, so that the remaining operator has to have a $(2\,2)$ part to it. If the coefficient of this extra operator is zero, the Hamiltonian in (31) represents the entire effect of the central interaction throughout the shell (or for the range of nuclei fitted).

The decomposition in (31) comes from a fit to the $A = 10$–12 nuclei and the coefficient of the extra $(2\,2)$ tensor is very small. Figure 3 shows the binding energies given by (31). The dashed lines show the energies for the full Hamiltonian in the cases of ^{10}Be, ^{11}B, and ^{12}C. The gain is about $4\,\mathrm{MeV}$ in each case and is mostly due to turning on the spin-orbit interaction. The circled numbers give the differences in the expectation value of the space-exchange component $(n_s - n_a)$ for successive spatial symmetries. Four times the coefficient of the space-exchange operator in (31) is $\sim 15.6\,\mathrm{MeV}$ which is very close to the energy of the first $T = 1$ states in ^{12}C. Thus, it is evident that the SU(4) invariant part of the central interaction gives a rather good account of the general structure of p-shell nuclei. The spin-orbit interaction, which transforms as $(1\,1)$ mixes spatial symmetries and L values. As will be seen, the interplay between the spin-orbit and tensor interactions can be very important and is quite subtle.

5.2 Structure of ^{6}Li, ^{7}Li, and ^{8}Be

The energy level schemes of these nuclei are shown in Fig. 4. All the levels shown can be accounted for by p-shell calculations. The lowest levels ($T = 0$ for ^{6}Li) have well-developed $\alpha + d$, $\alpha + t$, and $\alpha + \alpha$ cluster structures. For harmonic oscillator radial wave functions, coordinate transformations can be made on the states with maximal spatial symmetry so that all the quanta associated with the p-shell orbits reside on the relative coordinate between clusters formed from internal $0s$ wave functions (and the center of mass is in a $0s$ state). These states must transform as $(\lambda\,0)$ where λ is the number of quanta. Oscillator shell-model configurations beyond the p-shell are required to improve the radial behavior of the relative wave functions.

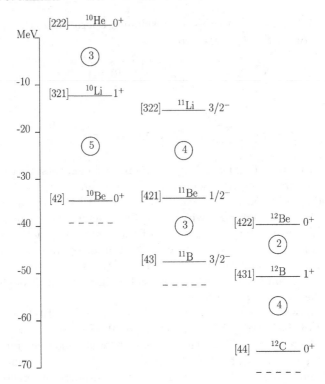

Fig. 3. Binding energy contributions from the SU(4) invariant part of the central interaction given by (31). The dashed lines give the result for the full p-shell Hamiltonian for the ground states of the nuclei with the highest spatial symmetries. The circled numbers are differences in $n_s - n_a$ for successive symmetries

Wave functions for ^{6}Li have been given in Table 5. It can be seen that LS coupling is rather good and that the $3^+; 0$, $2^+; 0$ and $1_2^+; 0$ states form a triplet with L=2, and S=1. Both vector and tensor forces can contribute to the splitting of this triplet. The most natural explanation is that the splitting is mainly due to the one-body spin-orbit interaction, partly because the even-state spin-orbit interaction acts in relative d states and the matrix elements are small in a G matrix derived from a realistic NN interaction. This is not necessarily the case for a fitted interaction. For example, the Cohen and Kurath interactions have small one-body spin-orbit terms and substantial even-state and antisymmetric spin-orbit terms that act in part like a one-body spin-orbit interaction with a strength that depends linearly on n and ensures that the p-hole states at A=15 are split by just over 6 MeV. The small quadrupole moment of ^{6}Li (experimentally $-0.082\,\mathrm{fm}^2$) provides a constraint on the balance of spin-orbit and tensor interactions. Writing

$$|^6\mathrm{Li}(gs)\rangle = \alpha\,^3S_1 + \beta\,^3D_1 + \gamma\,^1P_1 \tag{35}$$

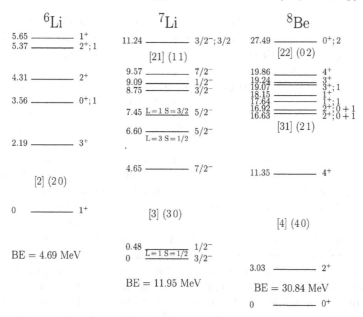

Fig. 4. Energy-level schemes for ^{6}Li, ^{7}Li, and ^{8}Be with the dominant spatial symmetry $[f]$, equivalently (λ, μ), indicated for groups of levels and L and S for particular levels. All energies are in MeV

leads to

$$Q(^6\text{Li}) = e^0 b^2 \left(\frac{4}{\sqrt{5}} \alpha\beta + \gamma^2 - \frac{7}{10}\beta^2 \right), \qquad (36)$$

where $b \sim 1.7$ fm is the oscillator parameter and $e^0 \sim 0.815$ is the isoscalar effective charge $(1 + \delta e_p + \delta e_n)/2$. The direct tensor interaction coupling the ^{3}S and ^{3}D states gives $\beta < 0$ while indirect coupling through the ^{1}P state via the spin-orbit interaction gives $\beta > 0$. Putting in numbers from Table 5 shows that a small negative value for β is required. The B(M1; 2^+;1 → 1^+;0) = $(8.3 \pm 1.5) \times 10^{-2}$ W.u. puts a similar restriction on β; briefly, the orbital contribution connecting ^{1}D to ^{1}P and the spin contribution connecting ^{3}P to ^{1}P are of the same sign while a spin contribution of the opposite sign from ^{1}D to ^{3}D cannot be too large if the B(M1) is to be reproduced. The interplay of spin-orbit and tensor interactions in leading to small but important wave function admixtures is a common feature in p-shell nuclei, most famously in the case of the very hindered ^{14}C(β^-) decay (see later).

In ^{7}Li, the lowest four states form the ground-state band and have > 93% purity of the indicated LS configurations. The first T = 3/2 has a similar purity of [2 1] symmetry with L = 1 and S = 1/2. An interesting point is that the second 5/2$^-$ state has a small width for decay into the α + t channel despite its proximity to the first 5/2$^-$ level which has a large decay width into this channel [29]. This means that the mixing matrix element between the [3] and lowest [21] symmetry $5/2^-$ states has to be small. Only a tensor interaction can connect the dominant components shown in Fig. 4 and this has to largely cancel with the spin-orbit contribution arising from a modest [21] ^{2}D component in the second state.

The hypernucleus $^8_\Lambda$Li (also $^8_\Lambda$Be) was frequently observed in emulsion studies [2] and analysis of the characteristic decay mode

$$^8_\Lambda \text{Li} \rightarrow \pi^- + {}^4\text{He} + {}^4\text{He} \tag{37}$$

established a ground-state spin-parity of 1^- and provided information on the mixing of configurations based on the ground-state and first-excited state of ^{7}Li. The configuration mixing is larger than usual because core states are close together and share the same L value. Both the ground-state spin of $^8_\Lambda$Li and the mixing provide restrictions on the nature of the ΛN effective interaction. To be studied by γ-ray spectroscopy, the A = 8 hypernuclei have to be formed by particle emission from a heavier hypernucleus.

The lowest 0^+ and 2^+ states of ^{8}Be form the core for bound states of $^9_\Lambda$Be (discussed in detail later). The 0^+ state is unbound by 92 keV and has a width of ~ 6 eV while the 2^+ state has a width of ~ 1.5 MeV. In the p-shell model, these states have very pure [4] symmetry with a few percent of [31] symmetry with S = 1. Because the Gamow-Teller operator cannot change spatial quantum numbers, it is these small admixtures in the 2^+ wave function that account for the β decays of ^{8}Li and ^{8}B. The near degeneracies of pairs of '[31]' states with the same J^π, different isospin, and similar space-spin wave functions lead to isospin mixing that is especially strong for the 16.63-MeV and 16.92-MeV 2^+ levels. The T = 1 analogs of these levels form the basis for the ground-state doublets of $^9_\Lambda$Li and $^9_\Lambda$B. The $^9_\Lambda$Li hypernucleus has been studied recently via the ^{9}Be (e, e'K$^+$) $^9_\Lambda$Li reaction [30].

The ground-state binding energies of the nuclei in Fig. 4 increase rapidly with the number of particles because the Pauli principle permits up to two neutrons and two protons to correlate strongly in spatially even states and take advantage of the strong central interaction in relative s states, as quantified in Sect. 5.1.

5.3 Structure of ^{9}Be and ^{11}C

Partial energy level schemes of ^{9}Be and ^{11}C are given in Fig. 5. These nuclei are paired together because, with p^5 and p^7 configurations, they are related by a particle-hole symmetry reflected in the conjugate SU(3) representations. Because the L values for the highest symmetry differ by steps of one (see Table 9), there are often two states with the same J value. To some extent, these states can be organized into K = 3/2 and K = 1/2 bands. In fact, it was shown a long time ago [31] that shell-model states for the p-shell nuclei have a large overlap with states angular momentum projected from a Slater determinant made from the lowest Nilsson model states (restricted to the

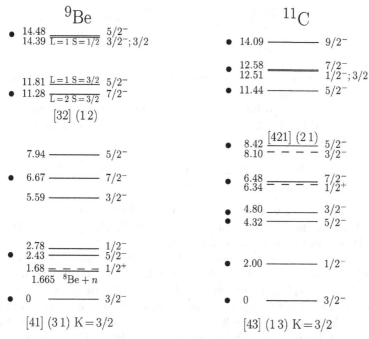

Fig. 5. Energy-level schemes for ^{9}Be and ^{11}C with the dominant spatial symmetry $[f]$, equivalently (λ, μ), indicated for groups of levels and L and S for particular levels. All energies are in MeV. The bullets mark levels strongly populated in proton knockout (or pickup) from ^{10}B and the ^{14}N(p, α)^{11}C reaction

p-shell and with the same one-body spin-orbit interaction as in the shell-model calculation) for some deformation. The deformations varied smoothly with the number of nucleons and were prolate at the beginning of the shell and oblate at the end of the shell. Thus, the lowest states of ^{6}Li, ^{7}Li, and ^{8}Be were obtained by filling the first K=1/2 Nilsson orbit. The K=3/2 orbit starts to fill at ^{9}Be and the second K=1/2 orbit is relatively close in energy. The ground-state of ^{11}C (or ^{11}B) would have three nucleons (or a hole) in the K=3/2 orbit.

The connection to SU(3) symmetry is quite close because Elliott [32] showed that all the angular momentum states for a given SU(3) representation could be projected out of a highest-weight state characterized by numbers of quanta $N_z = a + \lambda + \mu$ and $N_\perp = 2a + \mu$ with $K_L = \mu, \mu - 2, \ldots 1$ or 0

$$|(\lambda\mu)K_L LM\rangle = \frac{1}{a(K_L L)} P^L_{MK} \, \Phi(HW) \, . \tag{38}$$

The highest-weight state is made up of asymptotic Nilsson orbits (no spin-orbit interaction in this case). Something closer to reality can be obtained by projecting from a product of the highest-weight state and an intrinsic-spin wave function [33]

$$|(\lambda\mu)K_JLSJM\rangle = N\,P_{MK}^J\,\Phi(HW)\chi(SK_S)$$
$$= \sum_L c(L)|(\lambda\mu)K_LLSJM\rangle\,,\tag{39}$$

so that a given state with good K in general contains a mixture of L values. In SU(3) codes [34], the basis of (38) is used with the states orthogonalized with respect to K_L. The spin-orbit interaction can, and often does, mix L values to produce a good K_J. It also mixes $(\lambda\,\mu)$ and S values. For example, for the ^{11}C ground state,

$$|(1\,3)K=3/2\,J=3/2\rangle = \sqrt{21/26}\,|L=1\,S=1/2\rangle - \sqrt{5/26}\,|L=2\,S=1/2\rangle\,.\tag{40}$$

The CK816 interaction gives 0.7676 and -0.4833 for the coefficients, meaning that K$=3/2$ accounts for 81.3% out of a total 82.3% [43] symmetry. There is 13.5% [421] symmetry in the wave function.

An important point to notice is that the ^{9}Be ground state is not bound by much with respect to the neutron threshold (the ^{9}B ground state is unbound by 185 keV with respect to proton emission). This is an effect of the Pauli principle (embodied in the supermultiplet symmetry) which strongly restricts the way in which an extra p-shell nucleon can interact with a fully occupied orbit (in the Nilsson sense). On the other hand, the $1s\,0d$ states, which are near zero binding at this mass number, can couple to the ^{8}Be core without restriction. In fact, the low-lying positive-parity ($1\hbar\omega$) states also have a good SU(3) symmetry, namely (60) (typically $> 85\%$) obtained by coupling the (20) of the sd-shell nucleon to the (40) of the ^{8}Be core.

The levels of ^{9}Be in Fig. 5 marked by a bullet are strongly excited in proton knockout, or pickup, from ^{10}B [29]. The strength is governed by a spectroscopic factor which, by definition, is the square of the reduced matrix element of a creation operator connecting the two states involved. The J dependence of the reduced matrix element between basis states of the form (28) is contained in a normalized $9j$ symbol via (64). The reduced matrix element that remains is just $\sqrt{n}$ times a one-particle coefficient of fractional parentage (cfp) which defines how to construct a fully antisymmetric n-particle state from antisymmetric $(n-1)$-particle states coupled to the nth particle. Thus

$$\langle(\lambda\mu)\kappa LST\|a^+\|(\lambda'\mu')\kappa'L'S'T'\rangle$$
$$= \sqrt{n}\sqrt{\frac{n_{f'}}{n_f}}\langle(\lambda'\mu')\kappa'L'(1\,0)1\|(\lambda\mu)\kappa L\rangle\langle[\tilde{f'}]T'S'[\tilde{1}]1/2\,1/2\|[\tilde{f}]TS\rangle\,,\tag{41}$$

where the Clebsch-Gordan coefficients for SU(3) $\supset$ R(3) [34, 35] and SU(4) $\supset$ SU(2) $\times$ SU(2) [35, 36] result from applications of the Wigner-Eckart theorem for SU(3) and SU(4) and the weight factor $n_{f'}/n_f$ is the ratio of dimensions of representations of the symmetric groups S_{n-1} and S_n [35]. The $[f']$ are found by removing one box from the Young diagram for $[f]$ in all allowed ways. Examples of the weight factors for ^{9}Be and ^{10}B are given in Table 10.

Table 10. Weight factors for ^{9}Be and ^{10}B

^{9}Be $\rightarrow$ ^{8}Be + n	$\sqrt{n_{f'}/n_f}$	^{10}B $\rightarrow$ ^{9}Be + p	$\sqrt{n_{f'}/n_f}$
[41] $\rightarrow$ [4]	$\sqrt{\frac{1}{4}}$	[42] $\rightarrow$ [41]	$\sqrt{\frac{4}{9}}$
$\rightarrow$ [31]	$\sqrt{\frac{3}{4}}$	$\rightarrow$ [32]	$\sqrt{\frac{5}{9}}$

Because states with different supermultiplet symmetry are widely separated, the one-particle removal strength is in general complex [37]. The same is true for two-particle [38] and three-particle [39] removal but less so for the removal of an α particle [40] because the removed p^4 configuration is an SU(4) scalar. Pickup reactions provide a powerful way of identifying predominantly p-shell states. Stripping reactions are also very useful but can strongly populate states in which particles reside in higher shells (usually the next shell).

Table 10 shows that one reason why the binding energy of ^{9}Be is low with respect to ^{8}Be is that 3/4 of the parentage of the ^{9}Be ground state goes to highly-excited states of ^{8}Be and ^{8}Li. The weight factors for ^{10}B show that the parentage is almost equally divided between states of [41] and [32] symmetry. In fact, the lowest three states seen strongly in knockout are mainly [41] symmetry and the upper two states are mainly [32] symmetry. The upper $7/2^-$ state has a spectroscopic strength that is a factor of two larger than that for the lower $7/2^-$ state [29]. In the pure symmetry limit, this factor is ~ 7. The mixing of the two basis configurations needed to obtain the experimentally measured ratio is small. This is another case in which the balance between vector and tensor interactions in the mixing matrix element is important and different p-shell interactions tend to give rather different results for the ratio of strengths for the $7/2^-$ states.

A final observation for ^{9}Be is that the 11.81-MeV $5/2^-$ state is fed very strongly in the β^- decay of ^{9}Li [29] because it has largely the same spatial quantum numbers as the initial state. In fact, the B(GT) value is much larger than one would expect, perhaps because of difficulties in analyzing the $\alpha + \alpha + n$ final state. The analogous β^+ decay of ^{9}C [29] has close to the strength expected from shell-model calculations.

As expected, the ^{11}C (^{11}B) spectrum shows many similarities to the ^{9}Be spectrum. The positive-parity states are now more bound with respect to the nucleon threshold and, indeed, ^{11}Be has a $1/2^+$ ground state 0.32 MeV below the $1/2^-$ state. Because two particles can be promoted to the sd shell without breaking up the ^{8}Be core, $(sd)^2$ states are found quite low in energy, starting with the 8.10-MeV $3/2^-$ level. One-neutron removal from ^{12}C is limited to the first two $3/2^-$ states and the first $1/2^-$ state ([44] $\rightarrow$ [43] is unique). However, triton removal from ^{14}N via the ^{14}N(p, α)^{11}C reaction [41], and aided by the ^{3}D character of the ^{14}N ground state, strongly populates all the T = 1/2 p-shell states included in Fig. 5 (for theory, see [39]).

5.4 Structure of ^{10}B and ^{10}Be

Energy-level schemes of ^{10}B and ^{10}Be are given in Fig. 6. All the negative-parity states are shown. They are low in energy for the same reason that positive-parity states come low in ^{9}Be. Now, low-lying $(sd)^2$ states are possible because two p-shell nucleons that are strongly affected by the Pauli principle can be promoted to the sd shell without breaking up the [4] symmetry for the first four p-shell nucleons. Shell-model calculations show that all the states (except one) have the highest spatial symmetry and are dominated by the leading SU(3) symmetries, as indicated in Fig. 6.

The structure of the p-shell states of ^{10}B is interesting and is important for hypernuclear physics because ^{10}B forms the core for $^{11}_\Lambda$B which has been studied with the Hyperball detector. Six γ rays were observed but not all of them can be placed in a decay scheme. Even for those that can be placed with reasonable certainty, there are some puzzles (see later).

The (22) representation of SU(3) contains two L = 2 states [see Table 9 or (30)] and this is the only case in the p-shell for which the K_L quantum number is required. For ^{10}Be, S = 0 and the K assignments are clear and understandable in terms of two particles in the K = 3/2 Nilsson orbit or

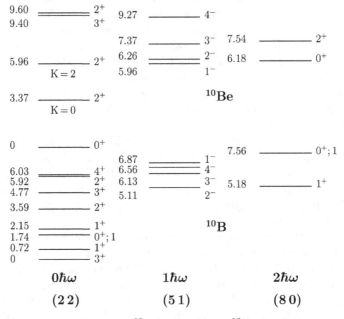

Fig. 6. Energy-level schemes for ^{10}B (bottom) and ^{10}Be (top). All energies are in MeV. All states have mainly [42] spatial symmetry except for the 9.60-MeV 2^+ level of ^{10}Be, which has mainly [33] symmetry. The neutron and α thresholds in ^{10}Be are at 6.812 MeV and 7.410 MeV. The α, deuteron, and proton thresholds in ^{10}B are at 4.461 MeV, 6.027 MeV, and 6.586 MeV

one each in the K = 3/2 and K = 1/2 orbits (these orbits have $K_L = 1$ and $K_S = \pm 1/2$). For ^{10}B, S = 1 and the $K_L = 0$ states are the 0.72-MeV 1^+ state with L = 0 and the L = 2 triplet of states at 2.15, 3.59, and 4.77 MeV. The ground state has K = 3, and mostly L = 2, and is connected by a very strong E2 transition to the 4^+ level at 6.03 MeV, there being a predicted but unobserved 5^+ level at higher energy. The 5.92-MeV 2^+ level is mainly L = 2 with $K_L = 2$ and in this sense is part of a triplet involving the 3^+ ground state and a 1^+ configuration predicted at higher energy. This triplet has the property of being very strongly split by the spin-orbit interaction while the $K_L = 0$ triplet remains much more compact. Electromagnetic transitions in ^{10}B have been investigated in great detail in the past [29] and it is from various selection rules that the K quantum numbers can be assigned. In particular, strong isovector M1 transitions must connect states with the same K_L.

5.5 Structure of ^{12}C, ^{13}C, and ^{14}N

The energy level schemes of ^{12}C, ^{13}C, and ^{14}N are given in Fig. 7. These nuclei are in a sense the particle-hole conjugates of the nuclei shown in Fig. 4. However, the effective spin-orbit interaction, indicated by the more than 6 MeV separation of the single-hole states of ^{15}N and ^{15}O, is much larger. The larger spin-orbit interaction tends to break the supermultiplet symmetry. Nevertheless, the content of the highest symmetry in the "ground-state" bands is typically > 70% and often higher.

There are now an increasing number of "intruder" levels marked by dashed lines. In ^{12}C, they include the Hoyle state at 7.65 MeV which is certainly not accounted for in shell-model calculations up to $2\hbar\omega$. The negative-parity states are, however, quite well accounted for in $1\hbar\omega$ shell-model calculations and have dominantly [44] symmetry and (3 3) SU(3) symmetry. The 0^+; 2 state is known to have a large, or even dominant, $(sd)^2$ component.

In ^{13}C, the extra p-shell nucleon is not well bound with respect to ^{12}C, the neutron threshold being at 4.95 MeV (cf. ^{9}Be vs. ^{8}Be) and positive-parity states, again unhindered by the Pauli principle, appear at low energies. The 8.86-MeV and 11.75-MeV levels are the lowest states with the [432] symmetry of the 15.11-MeV $3/2^-$ state, while the 9.90-MeV level is the lowest $(sd)^2$ state.

In ^{14}N, the lowest member of the marked group of predominantly ^{3}D two-hole states has become the ground state with the 3.95-MeV level being the predominantly ^{3}S state. The ground state is also predominantly two $p_{1/2}$ holes (there is an overlap of $\sqrt{20/27}$ with the ^{3}D configuration). The structure of the ^{14}N ground state is the important factor in the slowness of the ^{14}C β^- decay which is hindered by about six orders of magnitude compared with a strong allowed decay. Consider the following wave functions for the initial and final states in the the β^- decay

$$|^{14}C(0^+;1)\rangle = 0.7729\,^1S + 0.6346\,^3P$$
$$|^{14}N(1^+;0)\rangle = -0.1139\,^3S + 0.2405\,^1P - 0.9639\,^3D. \qquad (42)$$

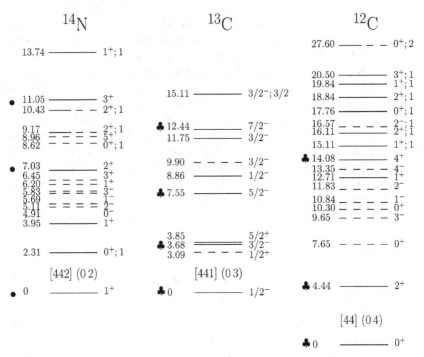

Fig. 7. Energy-level schemes for ^{12}C, ^{13}C, and ^{14}N. All energies are in MeV. For ^{12}C and ^{13}C, the club signs identify the members of the ground-state bands with the dominant symmetry indicated. For ^{14}N, the bullets indicate a triplet of states with ^{3}D two-hole configurations. Dashed lines indicate non p-shell states. The lowest particle thresholds are α at 7.367 MeV in ^{12}C, neutron at 4.946 MeV in ^{13}C, and proton at 7.551 MeV in ^{14}N

The Gamow-Teller matrix element is proportional to

$$\sqrt{3}a(^1S)\,a(^3S) + a(^1P)\,a(^3P) \tag{43}$$

and for the wave functions above the matrix element is $\simeq 0$. This is because the tensor interaction, essentially $\langle s|V_T|d\rangle$, was chosen to ensure the cancellation and kept fixed during a p-shell fit. This is another case where the spin-orbit interaction alone gives the wrong sign for the ^{3}S amplitude and a tensor interaction gives the opposite sign (see [42] for the history). Keeping the tensor interaction fixed leads to improvements in most of the cases for which the balance of tensor and vector interactions is important.

The above cancellation of the Gamow-Teller matrix element also plays an important role in the analogous M1 transition in ^{14}N. The absence of the normally dominant spin contribution to an isovector M1 transition leads to a rather small B(M1) dominated by the orbital contribution. This turns out to be important for understanding the properties of $^{15}_\Lambda$N which has been studied with the Hyperball.

Finally, the intruder positive-parity levels of ^{14}N shown in Fig. 7 are of $(sd)^2$ character, as one would expect from the presence of the low-energy positive-parity states in ^{13}C and ^{13}N, and in analogy to the $A = 10$ nuclei. The $2^+; 1$ levels have long been known to be of strongly mixed p-shell and $(sd)^2$ character.

6 The p-shell Hypernuclei

The structure of $^7_\Lambda$Li has already been discussed in Sect. 4 because, as an introduction to p-shell hypernuclei, it is a simple case with a p^2 ^{6}Li core that is amenable to hand calculation. This example was also used to compare and contrast the effects of Λ–Σ coupling in the s-shell and p-shell hypernuclei.

Following the survey of p-shell structure in terms of the LS-coupling supermultiplet basis in Sect. 5, this section is devoted to presenting the results obtained with the Hyperball on heavier p-shell hypernuclei and giving interpretations in terms of the underlying p-shell structure and effective YN interactions. The hypernuclei for which results have been obtained with the Hyperball in experiments at KEK and BNL are $^7_\Lambda$Li, $^9_\Lambda$Be, $^{10}_\Lambda$B, $^{11}_\Lambda$B, $^{12}_\Lambda$C, $^{15}_\Lambda$N, and $^{16}_\Lambda$O. For $^{16}_\Lambda$O, the calculation is a particle-hole calculation and for $^{15}_\Lambda$N, the calculation is similar to that for $^7_\Lambda$Li in that there are two p-shell holes instead of two p-shell particles.

6.1 The Shell-Model Calculations

The Hamiltonian

$$H = H_{\rm N} + H_{\rm Y} + V_{\rm NY} \,, \tag{44}$$

and the weak-coupling basis were introduced in (20) and (21). The formalism for the hypernuclear shell-model calculations is presented in Sect. 3.1 of [43] but some of the basic formulae are given here for completeness. The YN interaction can be written in terms of products of two creation and two annihilation operators with coefficients that are essentially the two-body matrix elements. The $a^+a^+a\,a$ product can be recoupled in any convenient order using any convenient coupling scheme. In the present case, it is convenient to write the operator in terms of a^+a pairs for the nucleons and hyperons so that we have a zero-coupled product of operators for separate spaces for which the matrix elements may be separated using the formulae in Appendix A. Formally,

$$V = \sum_\alpha C(\alpha) \left[\left[a^+_{j_{\rm N}} \widetilde{a}_{j_{\rm Y}} \right]^{J_\alpha T_\alpha} \left[a^+_{j'} \widetilde{a}_{j'_{\rm Y}} \right]^{J_\alpha T_\alpha} \right]^{00} \,, \tag{45}$$

where α stands for all the quantum numbers and the properly phased annihilation operators are given by

$$a_{jm\frac{1}{2}m_t} = (-)^{j-m+\frac{1}{2}-m_t} \widetilde{a}_{j-m\frac{1}{2}-m_t} \,, \tag{46}$$

and

$$
C(\alpha) = \sum_{KT} \begin{pmatrix} j_N & j_Y & K \\ j'_N & j'_Y & K \\ J_\alpha & J_\alpha & 0 \end{pmatrix} \begin{pmatrix} 1/2 & t_Y & T \\ 1/2 & t'_Y & T \\ T_\alpha & T_\alpha & 0 \end{pmatrix}
$$
$$
\times \widehat{K}\widehat{T} \langle j_N j_Y t_Y; KT \,|\, V \,|\, j'_N j'_Y t'_Y; KT \rangle . \tag{47}
$$

Then

$$
\langle \alpha_c J_c T_c, j_Y t_Y; JT \,|\, V_{NY-NY'} \,|\, \alpha'_c J'_c T'_c, j'_Y t'_Y; JT \rangle
$$
$$
= \sum_\alpha C(\alpha) \begin{pmatrix} J'_c & J_\alpha & J_c \\ j'_Y & J_\alpha & j_Y \\ J & 0 & J \end{pmatrix} \begin{pmatrix} T'_c & T_\alpha & T_c \\ t'_Y & T_\alpha & t_Y \\ T & 0 & T \end{pmatrix}
$$
$$
\times \frac{\widehat{J_\alpha T_\alpha}}{\widehat{j_Y t_Y}} \langle \alpha_c J_c T_c || (a^+_{j_N} \tilde{a}_{j'_N})^{J_\alpha T_\alpha} || \alpha'_c J'_c T'_c \rangle . \tag{48}
$$

The basic input from the p-shell calculation is thus a set of one-body density-matrix elements between all pairs of nuclear core states that are to be included in the hypernuclear shell-model calculation. As noted in Sect. 4, experimental energies are used for the diagonal core energies where possible.

The one-body transition density that governs the cross section for the formation of a particular hypernuclear state is (see Sect. 3.2 of [43])

$$
\langle p^{n-1}\alpha_c J_c T_f, j_\Lambda 0; J_f T_f || \left(a^+_{j_\Lambda} \tilde{a}_{j_N} \right)^{\Delta J 1/2} || p^n \alpha_i J_i T_i \rangle
$$
$$
= (-)^{j_N + j_\Lambda - \Delta J} U(J_i j_N J_f j_\Lambda, J_c \Delta J) \langle p^{n-1}\alpha_c J_c T_f || \tilde{a}_{j_N} || p^n \alpha_i J_i T_i \rangle . \tag{49}
$$

An important result is that in the weak-coupling limit the total strength for forming the states in a weak-coupling multiplet (summing over $J_f j_\Lambda$) is proportional to the pickup spectroscopic factor from the target [43]. To see the consequences of the spin-flip characteristics of the reaction used to produce the hypernuclear states, it is useful to change the coupling from $(j_N j_\Lambda)\Delta J$ to $(l_N l_\Lambda)\Delta L \Delta S \Delta J$ using (62).

6.2 The $^9_\Lambda$Be Hypernucleus

The bound-state spectrum for $^9_\Lambda$Be is shown in Fig. 8, which gives the γ-ray energies from an analysis of the BNL E930 data [44, 45], for the parameter set in (22) used for $^7_\Lambda$Li. An earlier experiment with NaI detectors [11] observed a γ ray at 3079(40) keV and put an upper limit of 100 keV on the doublet splitting.

The breakdown of the doublet splitting is given in Table 11. In the LS limit for ^{8}Be, the 2^+ wave function has L = 2 and S = 0. Then, only the coefficient of S_Λ survives and takes the value $-5/2$ as can be seen from an equation analogous to (11) with $\boldsymbol{S}_c$ replaced by $\boldsymbol{L}_c$. In the realistic case, the

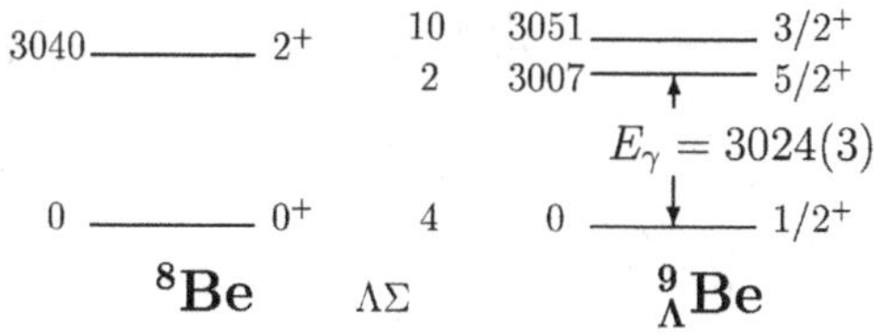

$$3040 \underline{\quad\quad} 2^+ \qquad \begin{matrix} 10 \\ 2 \end{matrix} \qquad \begin{matrix} 3051 \underline{\quad\quad} 3/2^+ \\ 3007 \underline{\quad\top\quad} 5/2^+ \end{matrix}$$

$$E_\gamma = 3024(3)$$

$$0 \underline{\quad\quad} 0^+ \qquad 4 \qquad 0 \underline{\quad\downarrow\quad} 1/2^+$$

$$^8\text{Be} \qquad \Lambda\Sigma \qquad ^9_\Lambda\text{Be}$$

Fig. 8. Energy levels of $^9_\Lambda$Be and the ^{8}Be core. The small shifts due to Λ–Σ coupling are shown in the center. All energies are in keV. The measured γ-ray energies are 3024(3) and 3067(3) keV giving a doublet separation of 43(5) keV [45]

contributions of S_Λ and T work against those from Δ and the Λ–Σ coupling (small in this case because the Σ has to be coupled to $T=1$ states of the core with a different symmetry from the $T=0$ states). A similar thing happens for the excited-state doublet of $^7_\Lambda$Li and the experimental results for both doublets restrict the combined effect of S_Λ and T to be small.

The parameter set chosen puts the $3/2^+$ state above the $5/2^+$ state but the order is not determined by this experiment. However, in the 2001 run of BNL E930 on a ^{10}B target, only the upper level is seen following proton emission from $^{10}_\Lambda$B. It can then be deduced that the $3/2^+$ state is the upper member of the doublet via the following reasoning. Four states of ^{9}B are strongly populated by neutron removal from ^{10}B [29] and the hypernuclear doublets based on these states are shown in Fig. 9. The structure factors which govern the population of these states are given at the right of the figure for two p-shell interactions. As discussed in Sect. 5.3, the relative neutron pickup strength to the two $7/2^-$ states which give rise to the $3^-/4^-$ doublets above the $^9_\Lambda$Be$^* + $p threshold is very sensitive to the non-central components of the p-shell interaction. Formation of the 3^- states is favored for the dominant $p_{3/2}$ removal by the coupling to get $\Delta L=1$ and $\Delta S=0$. The proton decay arises from ^{9}B$(7/2^-) \rightarrow {}^8$Be$(2^+) + $p in the core. The 4^- states proton decay to $^9_\Lambda$Be$(5/2^+)$ and from the recoupling $(2^+ \times p_{3/2})7/2^- \times s_\Lambda \rightarrow (2^+ \times s_\Lambda)J_f \times p_{3/2}$, governed by

$$(-)^{3/2+J_f-3} \, U(3/2\,2\,3\,1/2, 7/2\,J_f) \,, \tag{50}$$

Table 11. Contributions from Λ–Σ coupling and the spin-dependent components of the effective ΛN interaction to the $3/2^+$, $5/2^+$ doublet spacing in $^9_\Lambda$Be. The spectrum is shown on the right hand side of Fig. 8. As in Table 7, the first line gives the coefficient of each parameter and the second line gives the actual energy contributions in keV

$\Lambda\Sigma$	Δ	S_Λ	S_N	T	ΔE
	-0.033	-2.467	0.000	0.940	
-8	-14	37	0	28	44

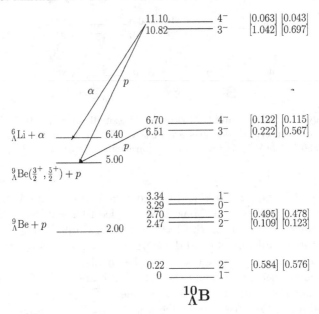

Fig. 9. Proton decay of $^{10}_\Lambda$B to $^9_\Lambda$Be. Formation strengths for non-spin flip production in the (K^-, π^-) reaction are given on the right for two p-shell models. Thresholds for particle decay of the $^{10}_\Lambda$B states are given on the left. All energies are in MeV

one finds that the 3^- states proton decay to the $3/2^+$ and $5/2^+$ states in the ratio of 32 to 3. Overall, the the $3/2^+$ state is favored by a factor of more than 3. The only caveat to this argument is that the uppermost 3^- state doesn't α decay too much.

6.3 The $^{16}_\Lambda$O Hypernucleus

At small angles in the $^{16}O(K^-, \pi^-)^{16}_\Lambda O$ reaction used for BNL E930, $p^{-1}p_\Lambda$ 0^+ states are strongly excited at about 10.6 and 17.0 MeV in excitation energy along with a broad distribution of $s^{-1}s_\Lambda$ strength centered near 25 MeV [46]. These levels can decay by proton emission (the threshold is at ~ 7.8 MeV) to $^{15}_\Lambda$N via $s^4 p^{10}(sd)s_\Lambda$ components in their wave functions. The low-lying states of $^{15}_\Lambda$N shown in Fig. 10 can be populated by s-wave or d-wave proton emission and higher energy negative-parity states by p-wave emission.

The cross section for the 0^+ states drops rapidly with increasing angle while the $\Delta L = 1$ angular distribution rises to a maximum near $10°$ [43]. The population of the excited 1^- state is optimized by selecting pion angles near this maximum. The aim of the experiment was to observe γ-rays from the excited 1^- state to both members of the ground-state doublet and thus measure the doublet splitting. The doublet splitting is of interest because it depends strongly on the tensor interaction. For a pure $p^{-1}_{1/2}s_\Lambda$ configuration, the combination of parameters governing the doublet splitting is [9]

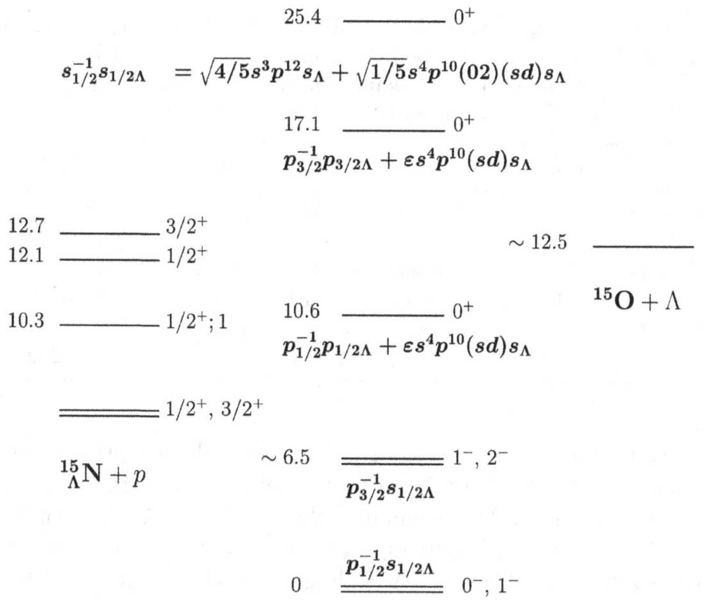

25.4 ———— 0^+

$$s_{1/2}^{-1}s_{1/2\Lambda} = \sqrt{4/5}s^3p^{12}s_\Lambda + \sqrt{1/5}s^4p^{10}(02)(sd)s_\Lambda$$

17.1 ———— 0^+

$$p_{3/2}^{-1}p_{3/2\Lambda} + \varepsilon s^4p^{10}(sd)s_\Lambda$$

12.7 ———— $3/2^+$

12.1 ———— $1/2^+$

~ 12.5 ————

10.3 ———— $1/2^+; 1$

10.6 ———— 0^+

$^{15}\mathrm{O} + \Lambda$

$$p_{1/2}^{-1}p_{1/2\Lambda} + \varepsilon s^4p^{10}(sd)s_\Lambda$$

═══ $1/2^+, 3/2^+$

$^{15}_\Lambda\mathrm{N} + p$

~ 6.5 ═══ $1^-, 2^-$

$$p_{3/2}^{-1}s_{1/2\Lambda}$$

0 ═══ $0^-, 1^-$

$$p_{1/2}^{-1}s_{1/2\Lambda}$$

Fig. 10. The energies of 1^- and 0^+ states of $^{16}_\Lambda\mathrm{O}$ that are strongly populated in the $^{16}\mathrm{O}(K^-, \pi^-)^{16}_\Lambda\mathrm{O}$ reaction [46] are shown in the center. All energies are in MeV. The dominant components of the wave function are shown together with the smaller admixtures that permit proton emission to states of $^{15}_\Lambda\mathrm{N}$

$$E(1_1^-) - E(0^-) = -\frac{1}{3}\Delta + \frac{4}{3}S_\Lambda + 8T . \tag{51}$$

The measured values of the γ-ray energies [47] are 6533.9 keV and 6560.3 keV (with errors of ~ 2 keV), giving 26.4 keV for the splitting of the ground-state doublet. Including recoil corrections of 1.4 keV to the γ-ray energies gives 6562 keV for the excitation energy of the 1^- state.

The breakdown of the contributions to the energy spacing in $^{16}_\Lambda\mathrm{O}$ from the shell-model calculation is given in Table 12 for the parameter set

$$\Delta = 0.430 \quad S_\Lambda = -0.015 \quad S_N = -0.350 \quad T = 0.0287 . \tag{52}$$

These were obtained by starting with the parameter values in (22) and changing T to fit the measured ground-state doublet spacing of $^{16}_\Lambda\mathrm{O}$ and S_N to fit the excitation energy of the excited 1^- level. The most important feature of the ground-state doublet splitting is the almost complete cancellation between substantial contributions from T and Δ (aided by Λ–Σ coupling). There is thus great sensitivity to the value of T if Δ is fixed from other doublet spacings.

Since [47] was published, another peak has been found at 6758 keV with a statistical significance of 3σ. The most likely interpretation is that it corresponds to the $2^- \to 1_1^-$ transition. The 2^- level has to be excited by a weak spin-flip transition and it is possible that states based on nearby levels

Table 12. Energy spacings in $^{16}_{\Lambda}$O. ΔE_C is the contribution of the core level spacing. The first line in each case gives the coefficients of each of the ΛN effective interaction parameters as they enter into the spacing while the second line gives the actual energy contributions to the spacing in keV

$J_i^\pi - J_f^\pi$	ΔE_C	$\Lambda\Sigma$	Δ	S_Λ	S_N	T	ΔE
$1^- - 0^-$			-0.380	1.376	-0.004	7.858	
	0	-30	-161	-21	1	226	27
$1^-_2 - 1^-_1$			-0.240	-1.252	-1.492	-0.720	
	6176	-30	-103	19	522	-21	6535
$2^- - 1^-_2$			0.619	1.376	-0.004	-1.740	
	0	81	266	-21	1	-50	292

of ^{15}O, shown in Fig. 11, could also be weakly excited. Accepting the first explanation puts the 2^- state at 6786 keV and implies a splitting of 224 keV for the excited-state doublet. This is smaller than the 292 keV given in Table 12 for value of Δ used for $^7_\Lambda$Li. Reducing Δ from 0.43 MeV to 0.33 MeV reduces the doublet splitting to 238 keV. A scaling of two-body matrix elements as $\sim A^{-0.3}$ is expected for heavier nuclei but for p-shell nuclei it is a more delicate question as could be anticipated from the discussion of Table 8. More evidence for a smaller value of Δ in the latter half of the p shell comes from doublet splittings in $^{15}_{\Lambda}$N and $^{11}_{\Lambda}$B.

6.4 The $^{15}_{\Lambda}$N Hypernucleus

As shown in Fig. 10, the high-energy 0^+ states of $^{16}_{\Lambda}$O populated strongly via the (K^-, π^-) reaction at forward pion angles (and 2^+ states at larger angles) populate states of $^{15}_{\Lambda}$N by proton emission. Three γ-ray transitions, corresponding to the solid arrows in Fig. 12 have been observed [1, 48]. The

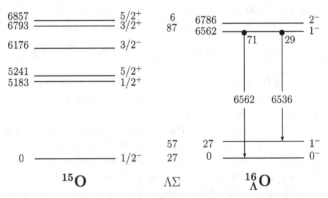

Fig. 11. Energy levels of $^{16}_{\Lambda}$O and the ^{15}O core. The shifts due to Λ–Σ coupling are shown in the center. All energies are in keV

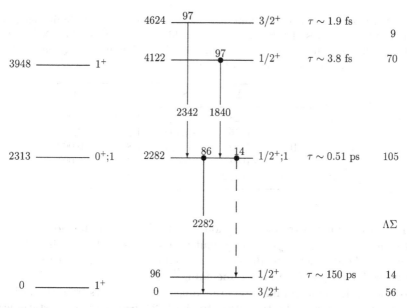

Fig. 12. The spectrum of $^{15}_\Lambda$N calculated from the parameters in (53). All energies are in keV. The levels of the ^{14}N core are shown on the left and the calculated lifetimes and shifts due to Λ–Σ coupling on the right

measured energies are 2268, 1961, and 2442 keV. The 2268-keV line is very sharp without Doppler correction, indicating a long lifetime compared to the stopping time in the target, and is identified with the transition from the $1/2^+$; 1 level to the $3/2^+$ member of the ground-state doublet. The other two γ-ray lines are very Doppler broadened and therefore associated with states that have short lifetimes.

The excited-state doublet splitting is calculated to be 637 keV with the parameter set (52). This is much larger than the observed spacing of 481 keV, much like the situation for the excited-state doublet of $^{16}_\Lambda$O. The results in Fig. 12, Table 13, and Table 14 are calculated with the parameter set

$$\Delta = 0.330 \quad S_\Lambda = -0.015 \quad S_N = -0.350 \quad T = 0.0239 , \tag{53}$$

where the value of T has been adjusted to fit the observed (26 keV) ground-state doublet spacing of $^{16}_\Lambda$O.

Table 13 shows the difference between the contributions of S_N for the mainly $p^{-2}_{1/2}$ and $p^{-1}_{1/2}p^{-1}_{3/2}$ core states. In LS coupling, the 1^+ ground state is mainly ^{3}D (42) and the excited 1^+ state is mainly ^{3}S. Looked at in this way, the coefficients of S_N for the last three states in the table arise mainly from the cross terms between the $L = 0$ and $L = 1$ components in the core wave functions. Small changes in the Λ–Σ coupling interaction can be used to fine tune the energy of the $1/2^+$; 1 state with respect to the $T=0$ states.

Table 13. Contributions of the spin-dependent ΛN terms to the binding energies of the five lowest states of $^{15}_{\Lambda}N$ given as the coefficients of each of the ΛN effective interaction parameters. In the $\Lambda\Sigma$ column, the gains in binding energy due to Λ–Σ coupling are given in keV (same as in Fig. 12)

$J_i^\pi;T$	$\Lambda\Sigma$	Δ	S_Λ	S_N	T
$3/2^+_1;0$	-56	-0.283	0.780	1.800	2.903
$1/2^+_1;0$	-14	0.457	-1.457	1.824	-6.053
$1/2^+_1;1$	-105	-0.022	0.021	1.816	-0.063
$1/2^+_2;0$	-70	-0.915	-0.084	0.447	0.091
$3/2^+_2;0$	-9	0.452	0.046	0.481	-0.333

The entries for the ground-state doublet of $^{15}_{\Lambda}N$ in Table 14 show a significant shift away from the jj-coupling limit with the result that the higher-spin member of the doublet is predicted to be the ground state in contrast to the usual case for p-shell hypernuclei, including $^{16}_{\Lambda}O$.

In the weak-coupling limit, the branching ratio for γ-rays from the $1/2^+;1$ state is 2:1 in favor of the transition to the $3/2^+$ final state (the statistical factor from the sum over final states). However, the transition to the $1/2^+$ state is not observed despite the fact that the transition to the $3/2^+$ state is very clearly observed with over 700 counts. In addition, a lifetime estimate for the $1/2^+;1$ level is 1.4 ps [48], which is very much longer than the 0.1 ps lifetime of $0^+;1$ level in ^{14}N. To understand these facts requires consideration of M1 transitions in ^{14}N and $^{15}_{\Lambda}N$ and this is the subject of the next subsection.

Table 14. Energy spacings in $^{15}_{\Lambda}N$. ΔE_C is the contribution of the core level spacing. The first line in each case gives the coefficients of each of the ΛN effective interaction parameters as they enter into the spacing while the second line gives the actual energy contributions to the spacing in keV. The first line of the table gives the coefficients for the ground-state doublet in the jj limit

$J_i^\pi - J_f^\pi$	ΔE_C	$\Lambda\Sigma$	Δ	S_Λ	S_N	T	ΔE
$p_{1/2}^{-2}$			0.5	-2.0	0	-12	
$1/2^+ - 3/2^+$			0.740	-2.237	0.024	-8.956	
	0	42	244	33	-8	-214	96
$1/2^+;1 - 3/2^+$			0.262	-0.752	0.016	-2.966	
	2313	-50	86	11	-5	-71	2282
$1/2^+_2 - 3/2^+_2$			1.367	0.130	0.034	-0.424	
	0	61	451	-2	-12	-10	502
$3/2^+_2 - 1/2^+;1$			0.474	0.025	-1.335	-0.271	
	1635	96	156	0	467	-6	2342

6.5 M1 transitions in ^{14}N and $^{15}_\Lambda$N

The effective M1 operator can be written

$$\boldsymbol{\mu} = g_l^{(0)}\boldsymbol{l} + g_l^{(1)}\boldsymbol{l}\tau_3 + g_s^{(0)}\boldsymbol{s} + g_s^{(1)}\boldsymbol{s}\tau_3 + g_p^{(0)}\boldsymbol{p} + g_p^{(1)}\boldsymbol{p}\tau_3 , \qquad (54)$$

where $\boldsymbol{p} = [Y^2, \boldsymbol{s}]^1$. The values of the effective g factors that fit the M1 properties of the single-hole states in ^{15}N and ^{15}O, and the states of interest in ^{14}N are given, along with the bare g factors, in Table 15.

The B(M1) value is given by

$$\text{B(M1)} = \frac{3}{4\pi}\frac{2J_f + 1}{2J_i + 1} \text{M}^2 , \qquad (55)$$

where

$$\text{M} = \langle f||\boldsymbol{\mu}^{(0)}||i\rangle + \langle T_i\, M_T\, 1\, 0\, |T_f\, M_T\rangle\langle f||\boldsymbol{\mu}^{(1)}||i\rangle . \qquad (56)$$

Contributions to the M1 matrix elements for ^{14}N M1 transitions are given in Table 16 for an interaction fitted in the manner described following (42) and (43). The important thing to notice, apart from the fact that the effective operator with this set of wave functions does describe the data well, is that the $0^+; 1 \to gs$ transition is hindered while the $1_2^+; 0 \to 0^+; 1$ transition is strong. In the former case, the $<\sigma\tau>$ matrix element is ~ 0 by construction (43) while in the latter it is very strong reflecting the allowed $^3S \to {}^1S$ nature of the transition. Also, the sign of the two matrix elements is different.

To see what this means for the M1 transitions de-exciting the $1/2^+; 1$ state of $^{15}_\Lambda$N, the most important components of the shell-model wave functions for $^{15}_\Lambda$N are listed in Table 17. The small $1_2^+; 0 \times s_\Lambda$ admixtures in the wave functions for the ground-state doublet will clearly lead to cancellations in the relevant M1 matrix elements because they bring in a large positive matrix element while the M1 matrix element between the large components is small and negative.

The general expression for electromagnetic matrix elements between hypernuclear basis states is

$$\langle (J_c T_c s_Y t_Y)\, J_f T_f ||M|| (J'_c T'_c s'_Y t'_Y)\, J_i T_i\rangle$$

$$= \delta_{YY'} \begin{pmatrix} J'_c & \Delta J & J_c \\ 1/2 & 0 & 1/2 \\ J_i & \Delta J & J_f \end{pmatrix} \begin{pmatrix} T'_c & \Delta T & T_c \\ t_Y & 0 & t_Y \\ T_i & \Delta T & T_f \end{pmatrix} \langle J_c T_c ||M_c^{\Delta J \Delta T}||J'_c T'_c\rangle$$

Table 15. Effective g factors for M1 transitions at the end of the p-shell. See [49] for theoretical estimates

	$g_l^{(0)}$	$g_s^{(0)}$	$g_p^{(0)}$	$g_l^{(1)}$	$g_s^{(1)}$	$g_p^{(1)}$
Bare	0.500	0.88	0	0.500	4.706	0
Effective	0.514	0.76	0	0.576	4.120	0.96

Table 16. Contributions to the M1 matrix elements for ^{14}N M1 transitions; μ is in μ_N and B(M1) is in W.u. (the M1 Weisskopf unit is $45/8\pi\ \mu_N^2$)

$J_f^\pi;T_f$	$J_i^\pi;T_i$	l or l_τ	s or s_τ	p or p_τ	$\mu/$B(M1)	Exp.	Bare g
$1_1^+;0$	$1_1^+;0$	0.7461	-0.3432	0	0.403	0.404	0.328
$1_1^+;0$	$0^+;1$	-0.5070	0.0003	0.2556	0.025	0.026(1)	0.077
$1_2^+;0$	$0^+;1$	-0.5590	3.4857	0.0304	3.50	3.0(9)	4.89
$1^+;0$	$2^+;1$	0.2282	-4.1491	0.1653	1.13	0.99	1.65
$2^+;0$	$2^+;1$	0.1651	3.7665	0.1884	2.26	2.29	2.64

$$+ \delta_{cc'} \begin{pmatrix} J_c & 0 & J_c \\ 1/2 & \Delta J & 1/2 \\ J_i & \Delta J & J_f \end{pmatrix} \begin{pmatrix} T_c & 0 & T_c \\ t_Y' & \Delta T & t_Y \\ T_i & \Delta T & T_f \end{pmatrix} \langle s_Y t_Y || M_Y^{\Delta J \Delta T} || s_Y' t_Y' \rangle. \quad (57)$$

The two important g factors in the hyperonic sector are $g_\Lambda = -1.226\,\mu_N$ and $g_{\Lambda\Sigma} = 3.22\,\mu_N$ (the g factors for the Σ hyperons are included in the calculations). For hypernuclear doublet transitions in the weak-coupling limit,

$$\mu = g_c \mathbf{J}_c + g_\Lambda \mathbf{J}_\Lambda$$
$$= g_c \mathbf{J} + (g_\Lambda - g_c)\mathbf{J}_\Lambda \quad (58)$$

can be used to obtain a simple expression for the matrix element in terms of $g_\Lambda - g_c$ as an overall multiplicative factor [9].

The important contributions for M1 decays from the $1/2^+;1$ state in $^{15}_\Lambda$N are shown in Table 18. The strong cancellation resulting from the small $1_2^+ \times s_\Lambda$ admixtures is evident. Even the small Σ admixtures contribute to the cancellation. The cancellation is stronger for the transition to the $1/2^+$ member of the ground-state doublet. The reason for this can be seen from Table 19. Namely, the largest contributions to the off-diagonal matrix elements come from S_N and T and add for the $1/2^+$ state and cancel for the $3/2^+$ state.

Finally, the M1 transition data for $^{16}_\Lambda$O and $^{15}_\Lambda$N are collected in Table 20, mainly to emphasize the weakness of the M1 transitions from the $1/2^+;1$ level of $^{15}_\Lambda$N.

Table 17. Excitation energies and weak-coupling wave functions for $^{15}_\Lambda$N

$J_n^\pi;T$	E_x (keV)	Wave function
$3/2_1^+;0$	0	$0.9985\ 1_1^+;0 \times s_\Lambda + 0.0318\ 1_2^+;0 \times s_\Lambda + 0.0378\ 2_1^+;0 \times s_\Lambda$
$1/2_1^+;0$	96	$0.9986\ 1_1^+;0 \times s_\Lambda + 0.0503\ 1_2^+;0 \times s_\Lambda$
$1/2_1^+;1$	2282	$0.9990\ 0_1^+;1 \times s_\Lambda + 0.0231\ 1_1^+;1 \times s_\Lambda + 0.0206\ 0_2^+;1 \times s_\Lambda$
		$-0.0261\ 0_1^+;1 \times s_\Sigma$
$1/2_2^+;0$	4122	$-0.0502\ 1_1^+;0 \times s_\Lambda + 0.9984\ 1_2^+;0 \times s_\Lambda$
$3/2_2^+;0$	4624	$-0.0333\ 1_1^+;0 \times s_\Lambda + 0.9984\ 1_2^+;0 \times s_\Lambda + 0.0363\ 2_1^+;0 \times s_\Lambda$

Table 18. Important contributions for M1 decays from the $1/2^+; 1$ state in $^{15}_\Lambda$N

$1/2^+; 1 \to 3/2^+; 0$	large component	$0.9979 \times 0.9988 \times (-0.251)$	-0.250
	$1^+_2 \times s_\Lambda$ admixture	$0.0318 \times 0.9988 \times (\ \ 2.957)$	$+0.095$
	Σ admixture		$+0.011$
	Partial sum		-0.137
$1/2^+; 1 \to 1/2^+; 0$	large component	$0.9983 \times 0.9988 \times (-0.251)$	-0.250
	$1^+_2 \times s_\Lambda$ admixture	$0.0545 \times 0.9988 \times (\ \ 2.957)$	$+0.161$
	Σ admixture		$+0.008$
	Partial sum		-0.081

6.6 The $^{10}_\Lambda$B, $^{12}_\Lambda$C, and $^{13}_\Lambda$C Hypernuclei

It was noted in Sect. 5.3 that ^{9}Be/^{9}B and ^{11}B/^{11}C have similar structure, as is evident from Fig. 5. The hypernuclei $^{10}_\Lambda$B and $^{12}_\Lambda$C will have $2^-/1^-$ ground-state doublets with 1^- as the ground state (this is known experimentally for $^{12}_\Lambda$B [2]). However, there are considerable differences in how these levels can be studied experimentally. In $^{10}_\Lambda$B, only the states of the ground-state doublet are particle-stable because, as Fig. 9 shows, the neutron threshold is at 2.00 MeV while the proton threshold is at 9.26 MeV in $^{12}_\Lambda$C. Fig. 9 also shows that the 2^- state of $^{10}_\Lambda$B is populated by non-spin-flip transitions from the 3^+ ground state of ^{10}B. The resulting γ-ray transition was first searched for in [13] without success, an upper limit of 100 keV being put on the doublet spacing (in BNL E930, the transition was also looked for and not found at roughly the same limit). In $^{12}_\Lambda$C, it is the 1^- ground state that is populated by non-spin-flip transitions from a ^{12}C target and the doublet spacing is best investigated by looking for transitions from higher bound states of $^{12}_\Lambda$C. This approach was tried in KEK E566 and the data is still under analysis.

The similarity of the contributions from the spin-dependent ΛN interaction to the two ground-state doublets is shown in Table 21 for a calculation using the parameters of (22) and the fitted p-shell interaction used for Fig 3. If the

Table 19. Coefficients of the ΛN interaction parameters in the off-diagonal matrix elements between the $1^+_1; 0 \times s_\Lambda$ and $1^+_2; 0 \times s_\Lambda$ basis states in $^{15}_\Lambda$N and the 1^- states in $^{16}_\Lambda$O. The second line gives the energy contributions in MeV

J^π	Δ	S_Λ	S_N	T	ME
$1/2^+$	0.1275	-0.1275	0.4581	-4.0664	
	0.0421	0.0019	-0.1603	-0.0972	-0.214
$3/2^+$	-0.0637	0.0637	0.4581	2.0332	
	-0.0210	-0.0010	-0.1603	0.0486	-0.134
1^-	0.4714	-0.4714	0.	1.4142	
	0.1556	0.0071	0.	0.0338	0.196

Table 20. M1 transition strengths in $^{16}_{\Lambda}$O and $^{15}_{\Lambda}$N

$J^{\pi}_f ; T_f$	$J^{\pi}_i ; T_i$	E_{γ} (keV)	B(M1) (W.u.)	γ branch (%)	lifetime
$0^-_1 ; 1/2$	$1^-_2 ; 1/2$	6562	0.336	72.5	0.24 fs
$1^-_1 ; 1/2$	$1^-_2 ; 1/2$	6535	0.129	27.5	
$0^-_1 ; 1/2$	$1^-_1 ; 1/2$	26	0.176	weak	10 ns
$3/2^+_1 ; 0$	$1/2^+ ; 1$	2268	4.55×10^{-3}	86	0.51 ps
$1/2^+_1 ; 0$	$1/2^+ ; 1$	2172	8.89×10^{-4}	14	
$3/2^+_1 ; 0$	$1/2^+_1 ; 0$	96	0.240	weak/γ	150 ps
$1/2^+ ; 1$	$3/2^+_2 ; 0$	2442	1.133	96.9	1.9 fs
$1/2^+ ; 1$	$1/2^+_2 ; 0$	1961	1.080	97.4	3.8 fs

parameters of (53) are used the ground-state doublet spacings for $^{10}_{\Lambda}$B and $^{12}_{\Lambda}$C drop to 121 keV and 150 keV, respectively.

The most notable point to be taken from Table 21 is that the Λ–Σ coupling increases the doublet spacing in $^{12}_{\Lambda}$C and reduces it in $^{10}_{\Lambda}$B. The reason for this is that spin-spin matrix element for the ΛN interaction depends on an isoscalar one-body density-matrix element of the nuclear spin operator for the core while the corresponding matrix element for Λ–Σ coupling depends on an isovector one-body density-matrix element of the nuclear spin operator for the core [see (48)]. The isoscalar and isovector matrix elements are both large but they have opposite relative sign for the two hypernuclei (this is a type of particle-hole symmetry for the K $= 3/2$ Nilsson orbit). The coupling matrix elements are broken down in Table 22. The "diagonal" matrix elements involving the $3/2^-$ core states contain a contribution of 1.45 MeV from $\overline{V}{\,}'$ (24) and the contribution from Δ' produces the shifts from this value. If it were not from the contribution to the energy shifts from the $1/2^- \times \Sigma$ configuration (the $1/2^-$ and $3/2^-$ core states both have L$=1$), there would be a much larger effect on the relative ground-state doublet spacings in $^{10}_{\Lambda}$B and $^{12}_{\Lambda}$C.

Apart from the effect of Λ–Σ coupling, several of the coefficients in Table 21 are sensitive to the model of the p-shell core. For example, the ground states of the core nuclei tend to be characterized by a good K value and this involves a mixing of L values as noted in, and following, (40). For L$=1$, the coefficient of Δ contributing to the doublet spacing is 2/3 whereas for L$=2$ the coefficient is $-2/5$. For the wave function in (40), the coefficient is the 6/23 $\sim$ 0.46. The

Table 21. Coefficients of the ΛN interaction parameters for the $2^-/1^-$ ground-state doublet separations of $^{10}_{\Lambda}$B and $^{12}_{\Lambda}$C. The energy contributions from Λ–Σ coupling and the doublet splitting ΔE are in keV

	$\Lambda\Sigma$	Δ	S_{Λ}	S_N	T	ΔE
$^{12}_{\Lambda}$C	58	0.540	1.44	0.046	-1.72	191
$^{10}_{\Lambda}$B	-15	0.578	1.41	0.013	-1.07	171

Table 22. Matrix elements (in MeV) coupling Σ configurations with the members of the $3/2^- \times \Lambda$ ground-state doublets in $^{10}_{\Lambda}$B and $^{12}_{\Lambda}$C. The energy shifts caused by these couplings are given in keV

J^π	$3/2^- \times \Sigma$	$1/2^- \times \Sigma$	$\Lambda\Sigma$ shift
$^{10}_{\Lambda}$B 1^-	0.55	1.47	34
2^-	1.95		49
$^{12}_{\Lambda}$C 1^-	1.92	-1.35	98
2^-	1.13		40

CK816 interaction gives a coefficient close to this value and the results of a calculation for $^{12}_{\Lambda}$C with this interaction and the parameter set (22) are shown in Fig. 13 and Table 23.

The non-observation of the ground-state doublet spacing in $^{10}_{\Lambda}$B is an important problem. A number of p-shell interactions give a smaller coefficient for Δ (due to more L mixing) or larger coefficient of T, both of which lead to a reduction in the doublet spacing. A better understanding of how the parameters vary with mass number and, indeed, whether the parametrization in use is sufficient are also important questions.

There have been many experiments using a ^{12}C target [1] and many show excitation strength in the region of the excited 1^- states. For example, in the (π^+, K^+) reaction with a thin carbon target, the second 1^- state is found at 2.5 MeV [4]. Table 23 shows that the S_N parameter is mainly responsible for raising the excitation energy above the core spacing of 2 MeV. Recently,

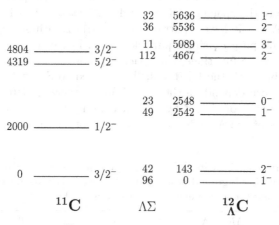

Fig. 13. The spectrum of $^{12}_{\Lambda}$C calculated from the parameters in (22). The levels of the ^{11}C core are shown on the left and the calculated shifts due to Λ–Σ in the center. All energies are in keV

Table 23. Energy spacings in $^{12}_\Lambda$C. $\Delta E_{\rm C}$ is the contribution of the core level spacing. The first line in each case gives the coefficients of each of the ΛN effective interaction parameters as they enter into the spacing while the second line gives the actual energy contributions to the spacing in keV

$J_i^\pi - J_f^\pi$	$\Delta E_{\rm C}$	$\Lambda\Sigma$	Δ	S_Λ	$S_{\rm N}$	T	ΔE
$2_1^- - 1_1^-$			0.463	1.518	0.030	−2.078	
	0	54	199	−23	−12	−62	143
$1_2^- - 1_1^-$			0.315	1.150	−1.104	0.635	
	2000	45	136	−17	430	19	2548
$1_3^- - 1_1^-$			0.372	−0.385	−1.647	0.561	
	4804	64	160	6	642	17	5536

excited states of $^{12}_\Lambda$B have been observed with better resolution via the $(e, e'K^+)$ reaction [50].

A similar effect of $S_{\rm N}$ is seen in Table 24 for $^{13}_\Lambda$C where the excited $3/2^+$ state built on the 4.44-MeV 2^+ state of ^{12}C (cf. Fig. 8) is seen at 4.880(20) MeV in a γ-ray experiment using NaI detectors [51] and at 4.85(7) MeV in KEK E336 via the (π^+, K^+) reaction [1]. Note that, as for $^9_\Lambda$Be, the effects of Λ–Σ coupling are small. In contrast to $^9_\Lambda$Be, the coefficients of $S_{\rm N}$ are large. This is because the ^{12}C core states, while still having dominantly [44] spatial symmetry, have substantial [431] components with S=1 (a low 68% [44] and 25% [431] in the ground state for the WBP interaction [52] used for Table 24 but typically $\sim$ 79% [44] for the Cohen and Kurath interactions).

The 4.88-MeV γ-ray was actually a by-product of an experiment [51] designed to measure the spacing of $1/2^-$ and $3/2^-$ states at $\sim$ 11 MeV in $^{13}_\Lambda$C ($B_\Lambda = 11.67$ MeV is the lowest particle threshold). To a first approximation, the two states are pure p_Λ single-particle states. In this case, and with harmonic oscillator wave functions, the spacing produced by the interaction of the p_Λ interacting with the filled s shell is related to S_Λ (with a coefficient of −6) because both depend on the same Talmi integral I_1 [10]. However, as noted above, the ^{12}C core is not by any means pure L=0, S=0 which means that components of the ΛN effective interaction other than the Λ spin-orbit

Table 24. Energy spacings in $^{13}_\Lambda$C. Coefficients of the ΛN effective interaction parameters for the $1/2^+$ ground state and $3/2^+$ excited state are given followed by the difference and the actual energy contributions in keV

J^π	$\Delta E_{\rm C}$	$\Lambda\Sigma$	Δ	S_Λ	$S_{\rm N}$	T	ΔE
$1/2^+$		27	−0.016	0.016	2.421	−0.049	
$3/2^+$		18	−0.045	−1.455	1.430	−0.929	
$3/2^+ - 1/2^+$	4439		−0.029	−1.471	−0.991	−0.880	
		9	−12	22	386	−28	4803

interaction, particularly the tensor interaction, play important roles in the small spacing of 152 keV [14]. In addition, the loose binding of the p_Λ orbit is important (the harmonic oscillator approximation is not good), as is configuration mixing produced by the quadrupole-quadrupole component in the $p_N p_\Lambda$ interaction [14, 43].

6.7 The $^{11}_\Lambda$B Hypernucleus

The $^{11}_\Lambda$B hypernucleus has a rather complex spectrum because the ^{10}B core has many low-lying p-shell levels, as shown in Fig. 6. The γ-decay properties of these levels have been very well studied [29]. Furthermore, the lowest particle threshold (proton) in $^{11}_\Lambda$B is at 7.72 MeV which means that the hypernuclear states based on the p-shell states of ^{10}B up to 6 MeV or so are expected to be particle stable and thus could be seen via their γ decay if they could be populated strongly enough.

A shell-model calculation for $^{11}_\Lambda$B was made using the p-shell interaction of Barker [53] who made some changes to one of the Cohen and Kurath interactions [25] to improve the description of electromagnetic transitions in ^{10}B. The strengths for formation via non-spin-flip transitions and the electromagnetic matrix elements for decay were calculated for all the bound p-shell hypernuclear states of $^{11}_\Lambda$B (i.e, up to the states based on the 5.92-MeV 2^+; 0 level of ^{10}B). The γ-ray cascade was followed from the highest levels, summing the direct formation strength and the feeding by γ rays from above. The conclusion was that perhaps as many as eight transitions would contain enough intensity to be seen in an experiment with the Hyperball. The formation strengths on the left side of Fig. 14 show that the most strongly formed excited state is expected to be the $3/2^+$ level based on the 5.16-MeV 2^+; 1 state of ^{10}B, followed by a number of states based on the low-lying 1^+; 0 and 0^+; 1 states. The lowest $1/2^+$; 0 level, originally predicted at 1.02 MeV, acts as a collection point for the γ-ray cascade. The predicted γ width at this energy corresponds to a lifetime of ~ 250 ps (the 1^+ state of ^{10}B has to decay by an E2 transition and has a lifetime of 1 ns) which is comparable with the expected lifetime for weak decay.

In the subsequent experiment, KEK E518, six γ rays were seen [1, 45, 54]. Figure 14 shows an attempt to construct a level scheme for $^{11}_\Lambda$B from a combination of the experimental information and the results of the shell-model calculation. The theoretical energies and the contributions from YN effective interactions are given in Table 25 for the parameter set (53). Apart from the γ-ray energies, the experimental information includes relative intensities and some estimate of the lifetimes from the degree of Doppler broadening.

The strongest γ ray in the spectrum was found at 1483 keV and it is very sharp implying a long lifetime. Despite the unexpectedly high energy, it is natural to associate this γ ray with the de-excitation of the lowest $1/2^+$; 0 level. The 2477-keV γ ray shows up after the Doppler-shift correction and it too has a natural assignment in Fig. 14. It is a 1.1 W.u. isovector M1 transition between

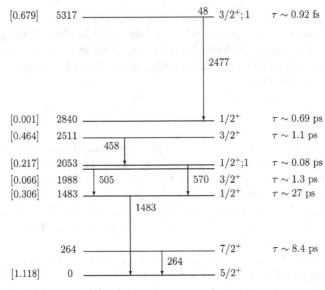

Fig. 14. The spectrum of $^{11}_\Lambda$B based on the six observed γ-ray transitions. All energies are in keV. The placements of the 264-keV, 1482-keV, and 2477-keV transitions are well founded. The placements of the other three γ-rays are more speculative. The formation factors for the (π^+, K^+) reaction on the left and the lifetimes on the right are from the shell-model calculation

states with L=2 and K_L=0. The 264-keV line is now known to be due to the ground-state doublet transition (0.2 W.u.), having been seen following proton emission from $^{12}_\Lambda$C [55] (this is the reason for showing a calculation using the parameter set (53) with $\Delta = 0.33$ MeV). The placement of the other three γ transitions in Fig. 14 is speculative, although the intensities and lifetimes

Table 25. Contributions of the spin-dependent ΛN terms to the binding energies of the eight levels of $^{11}_\Lambda$B shown in Fig. 14 given as the coefficients of each of the ΛN effective interaction parameters. The theoretical excitation energies and the gains in binding energy due to Λ–Σ coupling are given in keV

$J^\pi; T$	E_x	$\Lambda\Sigma$	Δ	S_Λ	S_N	T
$5/2^+; 0$	0	66	−0.616	−1.377	1.863	1.847
$7/2^+; 0$	266	11	0.409	1.090	1.890	−1.512
$1/2^+; 0$	968	71	−0.883	−0.116	0.746	0.243
$3/2^+; 0$	1442	12	0.403	0.094	0.872	−0.194
$1/2^+; 1$	1970	93	−0.007	0.008	1.543	−0.013
$3/2^+; 0$	2241	46	−0.266	0.754	1.536	−1.264
$1/2^+; 0$	2554	35	0.333	−1.333	1.674	2.639
$3/2^+; 1$	5366	103	−0.203	−1.293	1.519	0.598

match the theoretical estimates quite well. The $1/2^+; 1 \rightarrow 1/2^+; 0$ transition is 2.0 W.u. M1 transition between states with L = 0.

The most glaring discrepancy is that the shell-model calculation greatly underestimates the excitation energies of the two doublets based on the $1^+; 0$ levels of ^{10}B. From Table 25, it can be seen that S_N does raise the energies of these doublets with respect to the ground-state doublet but not nearly enough. The shell-model calculation is in fact quite volatile with respect to the p-shell wave functions for the $1^+; 0$ core levels. There is also mixing of the members of these two doublets and this is evident from the difference between the coefficients of S_N for the doublet members.

7 Summary and Outlook

The era of Hyperball experiments at KEK and BNL between 1998 and 2005 has provided accurate energies for about 20 γ-ray transitions in p-shell hypernuclei, the number in each hypernucleus being five for $^7_\Lambda Li$, two for $^9_\Lambda Be$, six for $^{11}_\Lambda B$, three for $^{15}_\Lambda N$, and three for $^{16}_\Lambda O$. Data from the last experiment, KEK 566, on a ^{12}C target using the upgraded Hyperball-2 detector array is still under analysis but there is evidence for one γ-ray transition in $^{12}_\Lambda C$ and two γ rays from $^{11}_\Lambda B$ have been seen following proton emission from the region of the p_Λ states of $^{12}_\Lambda C$ [55]. Several electromagnetic lifetimes have been measured by the Doppler shift attenuation method or lineshape analysis, and many estimates of, or limits on, lifetimes have been made based on the Doppler broadening of observed γ rays. In addition, two measurements of $\gamma \gamma$ coincidences have been made, for the $7/2^+ \rightarrow 5/2^+ \rightarrow 1/2^+$ cascade in $^7_\Lambda Li$ (471-keV and 2050-keV γ rays) and the $3/2^+ \rightarrow 1/2^+ \rightarrow 3/2^+$ cascade in $^{15}_\Lambda N$ (2442-keV and 2268-keV γ rays).

With the exception of transitions in $^{11}_\Lambda B$ that most likely involve levels based on the two lowest 1^+ states of ^{10}B, the γ-ray data can be accounted for by shell-model calculations that include both Λ and Σ configurations with p-shell cores. The spin-dependence of the effective ΛN interaction appears to be well determined. The singlet central interaction is more attractive than the triplet as evidenced by the value $\Delta = 0.43$ MeV needed to fit the 692-keV ground-state doublet separation in $^7_\Lambda Li$ (and the 471-keV excited-state doublet spacing). In $^7_\Lambda Li$, the contribution from Λ-Σ coupling is $\sim 12\%$ of the contribution from the ΛN spin-spin interaction in contrast to the 0^+, 1^+ spacings in the A = 4 hypernuclei, where the contributions are comparable in magnitude. The ΛN interaction parameters do exhibit a dependence on nuclear size. For example, the spacings of the excited-state doublets in $^{16}_\Lambda O$ $(1^-, 2^-)$ and $^{15}_\Lambda N$ $(1/2^+, 3/2^+)$, and the ground-state doublet in $^{11}_\Lambda B$ $(5/2^+, 7/2^+)$ require $\Delta \sim 0.32$ MeV. Given Δ, the tensor interaction strength T is well determined (~ 0.025 MeV) by the ground-state doublet $(0^-, 1^-)$ spacing in $^{16}_\Lambda O$ because of the sensitivity provided by a strong cancellation involving T and Δ. The Λ-spin-dependent spin-orbit strength S_Λ is constrained to be very small (~ -0.015 MeV) by the excited-state doublet spacing in $^9_\Lambda$Be ($3/2^+$, $5/2^+$). Finally, substantial effects of the nuclear-spin-dependent spin-orbit parameter $S_N \sim -0.4$ MeV, which effectively augments the nuclear spin-orbit interaction in changing the spacing of core levels in hypernuclei, are seen in almost all the hypernuclei studied. The small value of S_Λ and the substantial value for S_N mean that the effective LS and ALS interactions have to be of equal strength and opposite sign. The parametrization of the effective ΛN interaction includes some three-body effects (see later) but, if interpreted in terms of YN potential models, the value for Δ picks out NSC97e,f [5]. As noted in Sect. 3 and Sect. 4, these YN models have the correct combination of spin-spin and Λ–Σ coupling strengths to account for data on $^4_\Lambda$He ($^4_\Lambda$H) and $^7_\Lambda$Li. They also have weak odd-state tensor interactions that give a small positive value for T ~ 0.05 MeV. The LS interaction, which gives rise to $S_+ = (S_\Lambda + S_N)/2$, has roughly the correct strength but the ALS interaction is only about one third as strong as the LS interaction, although with the correct relative sign. For the newer ESC04 interactions [6], the ALS interaction is a little stronger and the other components seem comparable to those of the favored NSC97 interactions, except for differences in the odd-state central interaction. The attractive odd-state central interaction of the ESC04 models is favored by some data on p_Λ states over the overall repulsive interaction for the NSC97 models. The most recent quark-model baryon-baryon interactions of Fujiwara and collaborators [56] also have trouble explaining the small doublet splitting in $^9_\Lambda$Be.

The mass dependence of the interaction parameters has been studied by calculating the two-body matrix elements from YNG interactions using Woods-Saxon wavefunctions. This approach requires the assignment of binding energies for the p-shell nucleons and the s_Λ orbit. For the nucleons, binding energy effects are not so easy to deal with because, as emphasized in Sect. 5, the nuclear parentage is widely spread because of the underlying supermultiplet structure of p-shell nuclei (the allowed removal of a nucleon generally involves more than one symmetry [f] for the core and states with different symmetries are widely separated in energy). Perhaps the best that can be done is to take an average binding energy derived from the spectroscopic centroid energy for the removed nucleons. This changes rapidly for light systems up to ^{8}Be, beyond which it remains rather constant.

Any description of the absolute binding energies of p-shell hypernuclei (the B_Λ values) requires the consideration of binding-energy effects and the introduction of three-body interactions, real as in Fig. 1 or effective from many-body theory for a finite shell-model space. The empirical evidence for this need is that the B_Λ values for p-shell hypernuclei don't grow as fast as $n\overline{V}$, requiring a repulsive term quadratic in n (the number of p-shell nucleons) [10]. Also, a description of Λ single-particle energies over the whole periodic table requires that the single-particle potential have a density dependence [57], as might arise from the the zero-range three-body interaction in a Skyrme Hartree-Fock calculation. Much of the effect of a three-body interaction is included in the parametrization of the effective two-body ΛN interaction. If two or one s-shell nucleons are involved, the three-body interaction contributes to the Λ single-particle energy or the effective two-body ΛN interaction, respectively. This leaves only the $p^2 s_\Lambda$ terms to consider. The real three-body interactions derived from meson exchange present a problem for shell-model calculations in that they possess singular short-range behavior. In the two-body case, this is where the G matrix or a purely phenomenological treatment come in. For the three-body case, there are too many independent matrix elements to parametrize, although Gal, Soper, and Dalitz [7] have introduced a five parameter representation of the two-pion-exchange three-body interaction.

Given a set of three-body matrix elements, it is certainly possible to include them in the shell-model calculations [7]. Another useful extension of the shell-model codes would be to use the complete $1\hbar\omega$ space for the non-normal-parity levels of p-shell hypernuclei. Simple calculations for $p^n p_\Lambda$ configurations have been done [43] but is important to include the configurations involving $1\hbar\omega$ states of the core nucleus coupled to an s_Λ. This is necessary to permit the exclusion of spurious center-of-mass states from the shell-model basis and to provide the amplitudes for nucleon emission leaving low-lying states of the daughter hypernucleus with a Λ in the s_Λ orbit, as indicated in Fig. 10. The configuration mixing also redistributes the strength from from $p^n p_\Lambda$ states strongly formed in strangeness-exchange or associated-production reactions. In calculating matrix elements involving the p_Λ orbit, it is important to include binding-energy effects by using realistic radial wave functions because the p_Λ orbit becomes bound only at A $\sim$ 12 and the rms radius of the Λ orbit can be ~ 4.5 fm compared with ~ 2.8 fm for the p-shell nucleon.

The next generation of hypernuclear γ-ray spectroscopy experiments using a new Hyperball-J and the (K^-, π^-) reaction is being prepared for J-PARC, starting perhaps early in 2009. The day-one experiment [58] will be run at $p_K = 1.5$ GeV/c. The spin-flip amplitudes are strong in the elementary interaction between 1.1 GeV/c and 1.5 GeV/c and the cross sections for spin-flip vs non-spin-flip strength will be checked by using a ^{4}He target and monitoring the γ ray from the 1^+ excited state of $^4_\Lambda$He. Also, the intention is to make a precise measurement of the lifetime of the first-excited $3/2^+$ state of $^7_\Lambda$Li using the Doppler shift attenuation method. For $^{10}_\Lambda$B, the ground-state doublet spacing will be determined unless it is smaller than 50 keV. For $^{11}_\Lambda$B, the power of a larger and more efficient detector array will be used to sort out the complex level scheme by the use of $\gamma\gamma$ coincidence measurements. Finally, a ^{19}F target will be used to measure the ground-state doublet spacing in $^{19}_\Lambda$F.

The measurement on $^{19}_\Lambda$F represents the start of a program of γ-ray spectroscopy on sd-shell nuclei. This will require shell-model calculations for both $0\hbar\omega$ and $1\hbar\omega$ sd-shell hypernuclear states. In much of the first half of the sd shell, supermultiplet symmetry, SU(3) symmetry, and LS coupling are still rather good symmetries. As a result, there are the same opportunities as in the p shell to emphasize certain spin-dependent components of the effective

ΛN interaction by a judicious choice of target. Now there are more two-body matrix elements – 8 for $(sd)s_\Lambda$ – and more sensitivity to the range structure of the ΛN effective interaction.

The experiments just outlined represent the start of a very rich experimental program using Hyperball-J at J-PARC (see [1]). It should be possible to go to all the way to rather heavy nuclei where the p_Λ orbit is below the lowest particle-decay threshold (this is true for the special case of $^{13}_\Lambda C$ [51]).

This work has been supported by the US Department of Energy under Contract No. DE-AC02-98CH10886 with Brookhaven National Laboratory.

A Basics of Racah Algebra for SU(2)

The Wigner-Eckart theorem is used in the form [59]

$$\langle J_f M_f | T^{kq} | J_i M_i \rangle = \langle J_i M_i \, kq | J_f M_f \rangle \langle J_f || T^k || J_i \rangle \,. \tag{59}$$

The elements of the unitary transformation that defines the recoupling of three angular momenta are given by

$$| [(j_1 j_2) J_{12} j_3] \, J \rangle = \sum_{J_{23}} U(j_1 j_2 J j_3, J_{12} J_{23}) | [j_1 (j_2 j_3) J_{23}] \, J \rangle \,, \tag{60}$$

where the U-coefficient is simply related to the W-coefficient and 6j symbol [59]

$$U(abcd, ef) = \hat{e}\hat{f} W(abcd, ef) = \hat{e}\hat{f}(-)^{a+b+c+d} \begin{Bmatrix} a & b & e \\ d & c & f \end{Bmatrix} \,. \tag{61}$$

The elements of the unitary transformation that defines the recoupling of four angular momenta are given by

$$\begin{aligned} & | [(j_1 j_2) J_{12} (j_3 j_4) J_{34}] \, J \rangle \\ & = \sum_{J_{13} J_{24}} \begin{pmatrix} j_1 & j_2 & J_{12} \\ j_3 & j_4 & J_{34} \\ J_{13} & J_{24} & J \end{pmatrix} | [(j_1 j_3) J_{13} (j_2 j_4) J_{24}] \, J \rangle \,, \end{aligned} \tag{62}$$

where the normalized 9j symbol is simply related to the usual 9j symbol [59]

$$\begin{pmatrix} a & b & c \\ d & e & f \\ g & h & i \end{pmatrix} = \hat{c}\hat{f}\hat{g}\hat{h} \begin{Bmatrix} a & b & c \\ d & e & f \\ g & h & i \end{Bmatrix} \,. \tag{63}$$

The reduced matrix elements of a coupled operator consisting of spherical tensor operators that operate in different spaces, e.g. different shells or orbital and spin spaces, are given by

$$\langle J_1 J_2; J||[R^{k_1}, S^{k_2}]^k||J_1' J_2'; J'\rangle$$

$$= \begin{pmatrix} J_1' & k_1 & J_1 \\ J_2' & k_2 & J_2 \\ J' & k & J \end{pmatrix} \langle J_1||R^{k_1}||J_1'\rangle \langle J_2||S^{k_2}||J_2'\rangle . \tag{64}$$

The reduced matrix elements of a coupled operator consisting of spherical tensor operators that operate in the same space, e.g. a coupled product of creation and annihilation operators acting within the same shell, are given by

$$\langle x\Gamma||[R^\sigma, S^\lambda]^\nu||x'\Gamma'\rangle$$

$$= (-)^{\sigma+\Lambda-\nu} \sum_{y\Gamma_1} U(\Gamma\sigma\Gamma'\lambda, \Gamma_1\nu)\langle x\Gamma||R^\sigma||y\Gamma_1\rangle\langle y\Gamma_1||S^\lambda||x'\Gamma'\rangle , \tag{65}$$

where Γ represents all the angular momentum type quantum numbers such as JT or LST; x and y represent the other labels necessary to specify the states spanning a space.

B Two-body Matrix Elements of the ΛN Interaction

Here, the two-body $p_N s_\Lambda$ matrix elements of the ΛN effective interaction in (1) are given in both LS and jj coupling in terms of the parameters $\overline{V}$, Δ, S_Λ, S_N, and T from Table 2. Actually, the results are given in terms of S_+ and S_-, the radial matrix elements associated with the symmetric and antisymmetric spin-orbit interactions, respectively, so that $S_\Lambda = S_+ + S_-$ and $S_N = S_+ - S_-$. Note that in [7] the matrix elements are defined to give contributions to the Λ binding energy B_Λ and that the order of angular momentum coupling is (SL)J rather than (LS)J. In LS coupling,

$$\langle ^3P_0|V|^3P_0\rangle = \overline{V} + \frac{1}{4}\Delta - 2S_+ - 6T$$

$$\langle ^3P_1|V|^3P_1\rangle = \overline{V} + \frac{1}{4}\Delta - S_+ + 3T$$

$$\langle ^3P_2|V|^3P_2\rangle = \overline{V} + \frac{1}{4}\Delta + S_+ - \frac{3}{5}T$$

$$\langle ^1P_1|V|^1P_1\rangle = \overline{V} - \frac{3}{4}\Delta$$

$$\langle ^3P_1|V|^1P_1\rangle = -\sqrt{2}\,S_- . \tag{66}$$

In jj coupling,

$$\langle p_{3/2}\,1^-|V|p_{3/2}\,1^-\rangle = \overline{V} - \frac{5}{12}\Delta - \frac{1}{3}S_+ + T - \frac{4}{3}S_-$$

$$\langle p_{3/2}\,2^-|V|p_{3/2}\,2^-\rangle = \overline{V} + \frac{1}{4}\Delta + S_+ - \frac{3}{5}T$$

$$\langle p_{1/2}\,0^-|V|p_{1/2}\,0^-\rangle = \overline{V} + \frac{1}{4}\Delta - 2S_+ - 6T$$

$$\langle p_{1/2} \, 1^- | V | p_{1/2} \, 1^- \rangle = \overline{V} - \frac{1}{12}\Delta - \frac{2}{3}S_+ + 2\,T + \frac{4}{3}S_-$$

$$\langle p_{3/2} \, 1^- | V | p_{1/2} \, 1^- \rangle = \frac{\sqrt{2}}{3}\{\Delta - S_+ - S_- + 3\,T\} \ . \tag{67}$$

References

1. O. Hashimoto and H. Tamura: Prog. Part. Nucl. Phys. **57**, 564 (2006)
2. D. H. Davis and J. Pniewski: Contemp. Phys. **27**, 91 (1986)
3. M. Agnello et al: Phys. Lett. B **622**, 35 (2005)
4. H. Hotchi et al: Phys. Rev. C **64**, 044302 (2001)
5. T. A. Rijken, V. J. G. Stoks, and Y. Yamamoto: Phys. Rev. C **59**, 21 (1999)
6. T. A. Rijken and Y. Yamamoto: Phys. Rev. C **73**, 044008 (2006)
7. A. Gal, J. M. Soper, and R. H. Dalitz: Ann. Phys. (N.Y.) **63**, 53 (1971); ibid: **72**, 445 (1972); ibid: **113**, 79 (1978)
8. T. A. Brody and M. Moshinsky: *Tables of Transformation Brackets for Nuclear Shell-Model Calculations*, 2nd edn (Gordon and Breach, New York 1967)
9. R. H. Dalitz and A. Gal: Ann. Phys. (N.Y.) **116**, 167 (1978)
10. D. J. Millener, A. Gal, C. B. Dover, and R. H. Dalitz: Phys. Rev. C **31**, 499 (1985)
11. M. May et al: Phys. Rev. Lett. **51**, 2085 (1983)
12. V. N. Fetisov et al: Z. Phys. A **339**, 399 (1991)
13. R. E. Chrien et al: Phys. Rev. C **41**, 1062 (1990)
14. D. J. Millener: Nucl. Phys. A **691**, 93c (2001)
15. D. J. Millener: Nucl. Phys. A **754**, 48c (2005)
16. M. Bedjidian et al: Phys. Lett. B **83**, 252 (1979)
17. Y. Akaishi et al: Phys. Rev. Lett. **84**, 3539 (2000)
18. E. Hiyama et al: Phys. Rev. C **65**, 011301 (2001)
19. A. Nogga, H. Kamada, and W. Glöckle: Phys. Rev. Lett. **88**, 172501 (2002)
20. H. Nemura, Y. Akaishi, and Y. Suzuki: Phys. Rev. Lett. **89**, 142504 (2002)
21. H. Tamura et al: Phys. Rev. Lett. **84**, 5963 (2000)
22. K. Tanida et al: Phys. Rev. Lett. **86**, 1982 (2001)
23. M. Ukai et al: Phys. Rev. C **73**, 012501 (2006)
24. E. Hiyama et al: Phys. Rev. C **59**, 2351 (1999)
25. S. Cohen and D. Kurath: Nucl. Phys. **73**, 1 (1965)
26. J. Carlson: *LAMPF Workshop on* (π, K) *Physics*, Conference Proceedings No. 224, ed by B. F. Gibson, W. R. Gibbs, and M. B. Johnson (American Institute of Physics, New York, 1991) pp. 198–210
27. J. Carlson and R. Schiavilla: Rev. Mod. Phys. **70**, 743 (1998)
28. D. J. Millener and D. Kurath: Nucl. Phys. A **255**, 315 (1975)
29. http://www.tunl.duke.edu/nucldata/
30. Hall A collaboration, Thomas Jefferson Laboratory
31. D. Kurath and L. Pičman: Nucl. Phys. **10**, 313 (1959)
32. J. P. Elliott: Proc. Roy. Soc. A **245**, 128 (1958); ibid. p. 562
33. J. P. Elliott and C. E. Wilsdon: Proc. Roy. Soc. A **302**, 509 (1968)
34. J. P. Draayer and Y. Akiyama: J. Math. Phys. **14**, 1904 (1973)
35. H. A. Jahn and H. van Wieringen: Proc. Roy. Soc. A **209**, 502 (1951)
36. K. T. Hecht and S. C. Pang: J. Math. Phys. **10**, 1571 (1969)

37. S. Cohen and D. Kurath: Nucl. Phys. A **101**, 1 (1967)
38. S. Cohen and D. Kurath: Nucl. Phys. A **141**, 145 (1970)
39. D. Kurath and D. J. Millener: Nucl. Phys. A **238**, 269 (1975)
40. D. Kurath: Phys. Rev. C **7**, 1390 (1973)
41. C. C. Maples, LBL-253, U. Cal. thesis (1971), unpublished
42. H. J. Rose, O. Häusser, and E. K. Warburton: Rev. Mod. Phys. **40**, 591 (1968)
43. E. H. Auerbach et al: Ann. Phys. (N.Y.) **148**, 381 (1983)
44. H. Akikawa et al: Phys. Rev. Lett. **88**, 082501 (2002)
45. H. Tamura: Nucl. Phys. A **754**, 58c (2005)
46. W. Brückner et al: Phys. Lett. B **79**, 157 (1978)
47. M. Ukai et al: Phys. Rev. Lett. **93**, 232501 (2004)
48. M. Ukai et al: in preparation
49. I. S. Towner: Phys. Rep. **155**, 263 (1987)
50. M. Iodice et al: in preparation
51. H. Kohri et al: Phys. Rev. C **65**, 034607 (2002)
52. E. K. Warburton and B. A. Brown: Phys. Rev. C **46**, 923 (1992)
53. F. C. Barker: Aust. J. Phys. **34**, 7 (1981)
54. Y. Miura: Nucl. Phys. A **754**, 75c (2005); Ph.D Thesis, Tohoku University, 2005
55. H. Tamura: private communication
56. Y. Fujiwara et al: Phys. Rev. C **70**, 047002 (2004)
57. D. J. Millener, C. B. Dover, and A. Gal: Phys. Rev. C **38**, 2700 (1988)
58. J-PARC proposals, http://j-parc.jp
59. D. M. Brink and G. R. Satchler: *Angular Momentum*, 3rd edn (Oxford University Press, Oxford 1994)

Spectroscopy of Hypernuclei
with Meson Beams

Tomofumi Nagae

High Energy Accelerator Research Organization (KEK), J-PARC Project Office,
1-1 Oho, Tsukuba, Ibaraki 305-0801, Japan
tomofumi.nagae@kek.jp

1 Introduction

In a series of four lectures, I review the present status of experimental study
of hypernuclear spectroscopy by using $\pi^{\pm}$ and K^- beams. Further, I will give
an overview of the future spectroscopic studies at Japan Proton Accelerator
Research Complex (J-PARC), a high-intensity proton accelerator complex
now under construction in Japan [1].

2 Experimental Methods for Hypernuclear Spectroscopy

In the spectroscopy of hypernuclei, spectroscopic information we have obtained experimentally includes binding energies of the ground states of hypernuclei, energy levels of the excited states, and spin assignments of these states.
In order to resolve various energy levels, we need a good energy resolution in
hypernuclear mass measurements. For the spin assignment, unique features of
the reaction mechanism in production processes and/or decay processes have
been applied.

2.1 Mass Measurement

First, I discuss the hypernuclear mass measurement in the stopped-K^- reaction. A K^- beam, with rather low momenta, is injected on a nuclear target.
In some cases, thick targets are used to stop K^- before it decays (mean life
is $1.2385 \pm 0.0024 \times 10^{-8}$ s). K^- looses its energy in the target, and is eventually trapped in atomic orbits of a kaonic atom through various atomic processes. The K^- is absorbed in the final stage of atomic Auger process by
the atomic nucleus. In most cases ($\approx 80\%$), hyperons, Λ and Σ's, are produced via the $K^- + N \rightarrow \pi + Y\,(\Lambda$ or $\Sigma)$ reactions. The hyperons are produced with recoil momenta of $\approx 250\,\mathrm{MeV/c}$ for Λ and $\approx 180\,\mathrm{MeV/c}$ for Σ.

T. Nagae: *Spectroscopy of Hypernuclei with Meson Beams*, Lect. Notes Phys. **724**, 81–111
(2007)
DOI 10.1007/978-3-540-72039-3_3   © Springer-Verlag Berlin Heidelberg 2007

A small fraction (of $\approx 1\%$) of the produced hyperons is captured in the recoil nucleus to form a hypernucleus; namely a hypernucleus is formed in the $K^-_{stop} + A \rightarrow \pi + Hypernucleus$ reaction. Since this reaction is just a two-body process, the energy of the emitted pion tells us the mass of the produced hypernucleus as follows,

$$M_{\text{hyp}} = \sqrt{(E_\pi - M_{K^-} - M_A)^2 - \boldsymbol{p}_\pi^2} \,. \tag{1}$$

In this case, we have to measure the momentum of the emitted pion only. One magnetic spectrometer for π^- or a π^0 spectrometer detecting two gamma-rays from a π^0 decay with a good energy resolution is needed.

In the stopped-K^- reactions, there are other possibilities in which a Λ is captured by a nuclear fragment after loosing its energy by kicking out a few nucleons from the recoil nucleus, which is called a hyperfragment production. In this case, the identification of the hyperfragment species is not so easy. In a nuclear emulsion, we have a chance to observe almost all the emitted particles involved in the reactions so that the identification would be possible. The ground state of Λ hyperfragment decays through weak interaction. While non-mesonic weak decays ($\Lambda + N \rightarrow N + N$) are dominant in medium to heavy Λhypernuclei because of the Pauli blocking, mesonic weak decay has a substantial amount of decay branch in light hyperfragments. In some cases, it is known that the emitted pion energy is almost mono-energetic specifically indicating the binding energy of the hyperfragment. For example, the π^- momentum is 132.9 MeV/c in $^4_\Lambda H \rightarrow \pi^- + {}^4He$. A clear peak structure in the pion momentum is a good signal for the production of $^4_\Lambda H$.

In an experiment E906 of Brookhaven National Laboratory (BNL), a new method was applied to identify double-Λ hypernuclei by observing a sequential weak decays emitting two π^-'s with specific momenta. For example, in the case of $_{\Lambda\Lambda}^6 He \rightarrow {}^5_\Lambda He + p + \pi^- (T_\pi \approx 29.5$ MeV) and $^5_\Lambda He \rightarrow {}^4He + p + \pi^- (T_\pi \approx 31.7$ MeV), the two π^-'s have characteristic sharp peaks with the widths of $\Delta T_\pi \approx 0.10$ MeV and 1.0 MeV, respectively [2].

Secondly, direct reactions with meson beams have been used to produce various hypernuclei. The in-flight (K^-, π^-) reaction is a strangeness-exchanging reaction, and has a large cross section of the production of a hyperon (Y) of the order of mb/sr through the $K^- + N \rightarrow \pi + Y$ (Λ or Σ) reaction in the forward angle. This reaction is exothermic, so that the recoil laboratory momentum q of the hyperon is zero (recoilless) at the K^- incident momentum of ≈ 530 MeV/c and stays at ≤ 100 MeV/c in a wide momentum range as shown in Fig. 1.

In this case, the Λ hyperon has a rather large sticking probability to the recoil nucleus. Here the sticking probability is defined as

$$S_k(q; n_N l_N, n_Y l_Y) = |\langle \phi_{n_Y l_Y}(r) | j_k(qr) | \phi_{n_N l_N}(r) \rangle|^2 \,, \tag{2}$$

and is shown in Fig. 2 [3].

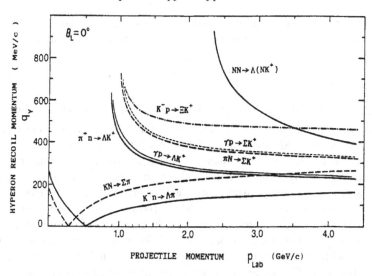

Fig. 1. The momentum q_Y transferred to the hyperon Y as a function of the projectile momentum $p_{Lab} = p_a$ in the reaction $aN \to Yb$ at $\theta_{b,L} = 0°$ [3]

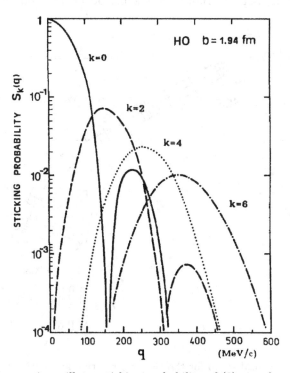

Fig. 2. The harmonic oscillator sticking probability of (2) as a function of q. The harmonic oscillator size parameter $b = 1.94$ fm is used [3]

The hypernuclear mass is obtained by measuring both the incident K^- momentum ($\boldsymbol{p}_{K^-}$) and the out-going π^- momentum ($\boldsymbol{p}_{\pi^-}$) as follows,

$$M_{\text{hyp}} = \sqrt{(E_{\pi^-} - E_{K^-} - M_A)^2 - (\boldsymbol{p}_{\pi^-} - \boldsymbol{p}_{K^-})^2} \, , \tag{3}$$

where E_{π^-} is calculated as $\sqrt{m_{\pi^-}^2 + p_{\pi^-}^2}$. Therefore, we need two magnetic spectrometer systems, and two systems must have a good resolution equivalently to achieve a good hypernuclear mass resolution (Fig. 3).

The (π^+, K^+) reactions have been also used for the production of Λ and Σ hypernuclei. The hypernuclear mass is obtained by measuring the π^+ momentum and the K^+ momentum with two spectrometers as in the case of the (K^-, π^-) reaction. Here, an $s\bar{s}$ pair is created from the vacuum and a K^+ and a Λ are produced in the final state (so called associated production). The production cross section for Λ peaks at the π^+ incident momentum at around $1.05\,\text{GeV/c}$ with the outgoing K^+ momentum at around $0.72\,\text{GeV/c}$ at the level of $\approx 10\,\mu\text{b/sr}$. Therefore, the central momenta of two spectrometers are higher than the case of the (K^-, π^-) reaction. The recoil momentum of the Λ is $\approx 350\,\text{MeV/c}$ as shown in Fig. 1.

In recent years, the $(e, e'K^+)$ reaction has been used for Λ hypernuclei spectroscopy at Hall-A and Hall-C in Thomas Jefferson Laboratory (JLab).

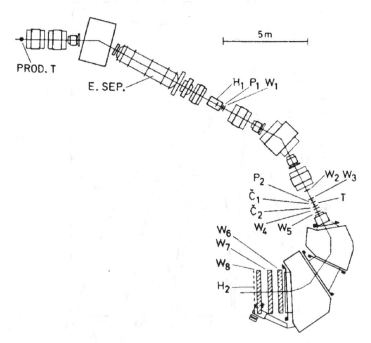

Fig. 3. An experimental setup for the (K^-, π^-) reaction at CERN. The label "T" indicates the experimental target position, and two magnetic spectrometers are installed both upstream and downstream of the target [4]

Since the primary electron beam at $\approx 1.8\,\mathrm{GeV}$ with $\Delta E/E \leq 1 \times 10^{-4}$ is used at Hall-C, a scattered electron ($\approx 0.3\,\mathrm{GeV}$) and a K^+ ($\approx 1.2\,\mathrm{GeV}/c$) must be momentum analyzed with two magnetic spectrometers. The recoil momentum of the Λ is at a similar level to the (π^+, K^+) reaction (Fig. 1). The production cross section by virtual photons is very small to be $\approx 50\,\mathrm{nb/sr}$. However the primary beam intensity of electrons ($\approx 30\,\mu\mathrm{A}$) compensates it.

2.2 Spin Assignment

Up to now, not so many ground-state spins of Λ hypernuclei have been known: $^{3}_{\Lambda}\mathrm{H}(1/2^+)$, $^{4}_{\Lambda}\mathrm{H}(0^+)$, $^{4}_{\Lambda}\mathrm{He}(0^+)$, $^{8}_{\Lambda}\mathrm{Li}(1^-)$, $^{11}_{\Lambda}\mathrm{B}(5/2^+)$, and $^{12}_{\Lambda}\mathrm{B}(1^-)$ [5]. These assignments were carried out with old emulsion data by using the reaction mechanism of the mesonic weak decay of Λ hypernuclei. Namely, in the mesonic weak decay, about 88% of the decay takes place via parity-violating (spin non-flip) amplitude. This is the dominance of the s-wave in the final state of $\Lambda \to \mathrm{p}\pi^-$ decay. Because of the interference between s-wave and p-wave amplitudes in the final state, the angular distribution of π^- emission has strong correlation to the axis of the Λ spin.

For example, in Fig. 4, the π^- decay angular distribution of $^{4}_{\Lambda}\mathrm{H}$, which is produced via the $K^-_{\mathrm{stop}} + {}^4\mathrm{He} \to {}^{4}_{\Lambda}\mathrm{H} + \pi^0$ reaction, is shown [6]. The

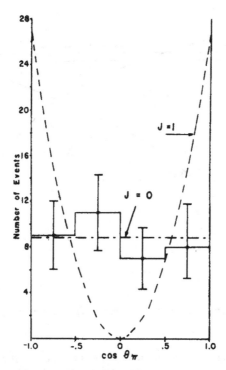

Fig. 4. The angular distribution of the π^- from the decay of $^{4}_{\Lambda}\mathrm{H} \to \pi^- + {}^4\mathrm{He}$ for hyperfragments produced in the capture reaction $K^-_{\mathrm{stop}} + {}^4\mathrm{He} \to {}^{4}_{\Lambda}\mathrm{H} + \pi^0$ [6]

distribution should be isotropic when the spin (J) of $^4_\Lambda$H is 0, while it should be $\cos^2\theta$ in the case of $J = 1$. Therefore, the isotropic distribution in Fig. 4 indicates $J = 0$.

An independent determination of the $^4_\Lambda$H spin was obtained from the π^- decay branching ratio based on an impulse model calculation of Dalitz and Liu [7]. Because of the dominance of the spin non-flip amplitude, the two-body decay of $^4_\Lambda$H$\rightarrow \pi^- +^4$He is suppressed when $J = 1$. Figure 5 shows the variation of

$$R_4 = \frac{\Gamma(^4_\Lambda\text{H} \rightarrow \pi^- +^4 \text{He})}{\Gamma(\text{all } \pi^- \text{ decays of } ^4_\Lambda\text{H})} \tag{4}$$

as a function of the relative s-wave and p-wave amplitudes in free hyperon decay for two spin possibilities $J = 0$ and $J = 1$ (shown as S in the figure). The experimental values of $R_4 = 0.69 \pm 0.02$ and $p^2/(s^2 + p^2) = 0.126 \pm 0.006$ are shown in the figure. It also supports $J = 0$.

In all cases known so far, the ground-state spin is expressed as $J_{\text{hyp}} = J_{\text{core}} - 1/2$, which indicates that the Λ-nucleon spin-spin interaction in spin-singlet state is more attractive than that of spin-triplet state.

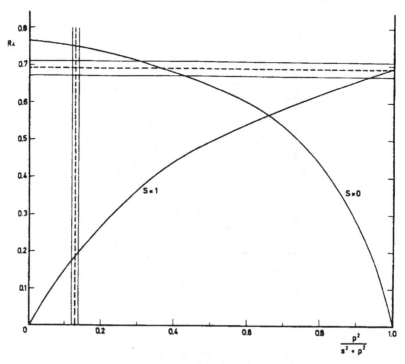

Fig. 5. The branching ratio R_4 as a function of $p^2/(s^2 + p^2)$ for the two spin possibilities 0 and 1 for $^4_\Lambda$H. The experimental values and their errors are shown [8]

2.3 Angular Distribution

The angular distribution of the forward cross section in the (K^-, π^-) reaction contains useful information on the angular momentum transfer. As shown in Fig. 2, when the recoil momentum is small at zero degrees, the angular momentum transfer $\Delta\ell$ (k in the figure)$=0$ is favored. Thus, so-called substitutional states, $(p_{3/2,n}^{-1}, p_{3/2,\Lambda})$, (s_n^{-1}, s_Λ), etc., are preferentially populated in the very forward angle. In contrast, when we go to a larger scattering angle, the recoil momentum is increased and the $\Delta\ell \geq 1$ transition becomes comparable to the $\Delta\ell = 0$ transition. Therefore, we would observe other transition peaks corresponding to the $(p_{3/2,n}^{-1}, s_{1/2,\Lambda})$, $(p_{3/2,n}^{-1}, p_{1/2,\Lambda})$, etc.

This specific feature of the angular distribution was well demonstrated at CERN as shown in Fig. 6.

On the other hand, the angular distributions obtained in the (π^+, K^+) reaction [10] and $(e, e' K^+)$ reaction are not expected to show this kind of specific feature for different transitions in the forward angles.

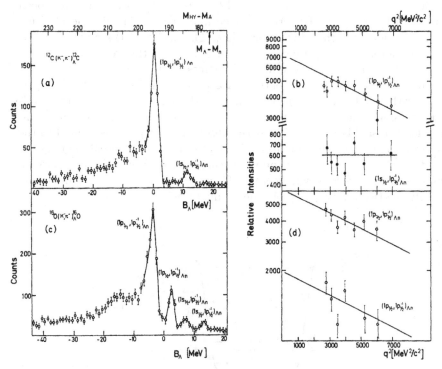

Fig. 6. Excitation spectra obtained from the (K^-, π^-) reactions on ^{12}C and ^{16}O at $p_K = 715$ MeV/c. The q-dependence (angular distribution) of the peak intensities is displayed in the right-half [9]

2.4 Hypernuclear γ-ray Spectroscopy

Measurements of γ-ray transitions in Λ hypernuclei enable us to resolve various excited levels with an excellent energy resolution. The energy resolution of ≈ 2 keV (≈ 100 keV) is achieved with Ge (NaI) detectors, while that of magnetic spectrometers are still limited at ≈ 700 keV$_{\mathrm{FWHM}}$ at this moment.

Nevertheless, there have been a lot of difficulties to apply γ-ray spectroscopy to hypernuclear spectroscopy. First, the detection efficiency of γ-ray measurements is not high enough and we thus need a large yield of hypernuclear productions, which has been limited by limited beam intensities so far. Also, it is preferable to cover a large solid angle with γ-ray detectors. However, the expensive cost of those detectors is a problem to overcome. Further, it has been not so easy to operate γ-ray detectors in such a high-rate environment with secondary meson beams from the technical viewpoint.

These issues have been solved somehow by Tamura [10] in constructing a large-acceptance germanium (Ge) detector array dedicated to hypernuclear γ-ray spectroscopy called Hyperball. With the first successful run at KEK 12-GeV Proton Synchrotron (PS) [11, 12], a series of experiments have been performed by using the Hyperball, which have revealed detailed level structures of various p-shell Λ hypernuclei [10]. The details of these hypernuclear γ-ray measurements are discussed by D.J. Millener in this volume.

However, there exist several weak points in hypernuclear γ-ray spectroscopy. A number of single-particle Λ orbits are bound in heavy Λ hypernuclei with a potential depth of around 30 MeV. However the energy levels of many single-particle orbits are above the nucleon (proton and neutron) emission thresholds (Fig. 7). Thus, the observation of γ rays is limited to the low excitation region, maybe up to the Λ p orbit. Another weak point is the fact that the γ-ray transition only measures the energy difference between two states. Therefore, single energy information only is not enough to fully identify the two levels; γ − γ coincidence might help to resolve it.

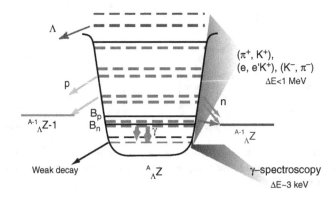

Fig. 7. Various decay schemes of Λ hypernuclei

2.5 Meson Beams

Meson beams, $\pi^\pm, K^\pm$, in the medium-momentum range of 1–2 GeV/c are produced in high-energy proton-nucleus reactions in the incident energy range of 10–50 GeV (Fig. 8). The production cross sections have a peak in the forward angles.

Because of the light mass 140 MeV/c^2, pions are abundantly produced with the proton beam at energies of more than a few GeV. Charge conservation prefers to produce positive pions rather than negative pions. Kaons have a heavier mass 500 MeV/c^2, so that the K/π ratio at the production stage is about 1/100 or less. Similar to the (π^+, K^+) reaction, in the proton-nucleus reactions kaons are produced associated with another strange particle. Thereby, strangeness is conserved. In the case of K$^+$, the production threshold is the $p + N \to K^+ + \Lambda + N$ threshold, while in the case of K$^-$ it is the $p + N \to K^+ + K^- + p + N$ threshold. Therefore, K$^-$ yield is less than K$^+$ yield (Fig. 9).

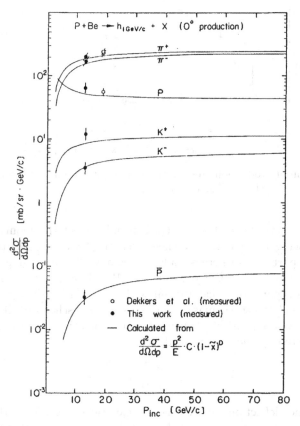

Fig. 8. Incident proton momentum dependence of the production of secondary beams at 1 GeV/c [13]

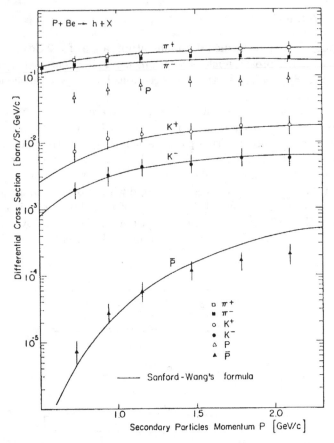

Fig. 9. Differential production cross sections of various secondary beams from the 12-GeV proton incident on a ^{9}Be target [13]

Since a magnetic beam line system can only select the momentum and charge of the secondary particles, we need an extra device to select the mass of the secondary particles. Otherwise, a kaon beam would be nothing but a pion beam with a very small fraction of kaons ($\approx$ 0.1%). In a low-energy kaon beam line, this mass separation is carried out with an electrostatic separator. The principle of the electrostatic mass separator system is schematically shown in Fig. 10. When charged particles traverse the electrostatic field perpendicularly, they are deflected by the electric field as

$$y' = \frac{e\,E\,\ell}{p\,\beta\,c}\,,\tag{5}$$

where y' is the deflection angle, e is the electric charge, E is the electric field, ℓ is the field length along the beam axis, p is the particle momentum, and β is the particle velocity. In most of the mass-separated beam lines, the

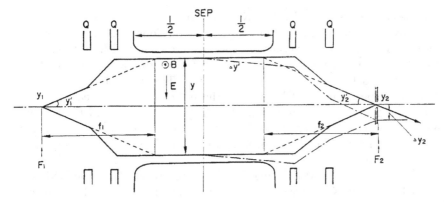

Fig. 10. Principle of electrostatic separator [13]

vertical angular-deflection y' for the wanted particle is cancelled by small crossed magnetic field $B\,(=E/\beta c)$ perpendicular to the vertical electric field to keep the trajectory for the wanted particle straight along the beam axis. The deviation of the deflection angle of the unwanted particle from the central beam orbit can be expressed as

$$\Delta y' = \frac{e\,E\,\ell}{p\,c}\left(\frac{1}{\beta_{\mathrm{w}}} - \frac{1}{\beta_{\mathrm{u}}}\right)\,, \tag{6}$$

where β_{w} and β_{u} are the velocities of the wanted and unwanted particle, respectively. The momentum p of both particles is the same. The angular separation $\Delta y'$ in the electrostatic separation is converted to the spatial separation Δy_2 by using a subsequent optical system, which consists of magnetic lenses and drift spaces (Fig. 10). A mass slit is introduced in the vertical direction to eliminate the unwanted particles for the actual mass separation.

Because of the limitation of the maximum electric field in the electrostatic separators, it is difficult to realize the $\mathrm{K^-/\pi^-}$ ratio of ≈ 1 with a single-stage electrostatic separator system. While with a double-stage system, the $\mathrm{K^-/\pi^-}$ ratio ≥ 1 is achieved at D6 beam line of BNL-AGS [14].

3 Brief History of Hypernuclear Spectroscopy

A hyperfragment that included a Λ hyperon was observed in 1952 [15] soon after the discovery of the Λ hyperon. Since then, there has been much progress in hypernuclear spectroscopy. The objects to be explored have been extended from Λ hypernuclei to various other hypernuclei, such as Σ hypernuclei, double-Λ hypernuclei, and Ξ hypernuclei. As for spectroscopic information, however, the number of states for which spin and parity have been experimentally established has been very limited so far.

In the early days around 1960s, K^- beam produced with accelerators was used for hyperfragments production in nuclear emulsions with the stopped-K^- reaction. The produced hypernuclear species were identified from their weak decay. Therefore, the information was restricted to ground states, mostly of light Λ hypernuclei. In 1970s, the so-called recoilless method was utilized in a series of experiments at CERN with the in-flight (K^-, π^-) reaction. The spectroscopic information was obtained with the missing-mass method. Thus, the information for the excited states of hypernuclei was extracted for the first time. For example, a small spin-orbit splitting was suggested for the p orbit of Λ hypernuclei.

Since the 1980s, the Alternating-Gradient Synchrotron (AGS) at BNL and the 12-GeV PS at KEK have played very important roles in Hypernuclear Physics by providing high-intensity K^- and $\pi^\pm$ beams. The (π^+, K^+) and (K^-, π^-) reactions were used to produce Λ hypernuclei and Σ hypernuclei, and the (K^-, K^+) reaction to investigate strangeness $S = -2$ systems, such as double-Λ hypernuclei, Ξ hypernuclei, and the hypothetical H dibaryon. The D6 beam line at BNL-AGS played a unique and important role in studies of $S = -2$ systems, providing a high-quality K^- beam at 1.8 GeV/c [16]. At KEK-PS, the SKS spectrometer was essential to conduct Japanese hypernuclear programs. The Hyperball detector has opened a new regime of hypernuclear spectroscopy in precision. These experiments at BNL-AGS and KEK-PS have extended the scope of hypernuclear physics into the physics of hadron many-body systems with strangeness degrees-of-freedom, i.e. Strangeness Nuclear Physics.

In 2000, Thomas Jefferson Laboratory (JLab) started hypernuclear spectroscopy with a different reaction of $(e, e'K^+)$ [10, 17]. The first measurement demonstrated a good energy resolution of 0.7 MeV. In 2004, the FINUDA experiment at Frascati also completed its first run for (K^-_{stop}, π^-) spectroscopy with 1.3-MeV resolution [18].

4 $(\pi^\pm, K^+)$ Spectroscopy

The spectroscopy of Λ hypernuclei has been successfully carried out at KEK-PS using the SKS spectrometer [19]. An overview of the measurements with the (π^+, K^+) reactions is given in [10, 20].

The SKS spectrometer (Fig. 11) is installed at the K6 beam line of the KEK 12-GeV proton synchrotron, where a mass-separated pion beam is available with a typical beam intensity of $\approx 2 \times 10^6$ every 3 seconds at about 1 GeV/c. It has been used for various kinds of experiments in intermediate-energy nuclear physics: the spectroscopy of Λ hypernuclei via the (π^+, K^+) reaction, weak decays of Λ hypernuclei, pion-nucleus reactions, etc.

The specific feature of the SKS is that it has a good energy resolution of $\leq 2\,\text{MeV}_{\text{FWHM}}$ and a large solid angle of 100 msr. A large superconducting dipole magnet with a pole gap of 50 cm and the maximum magnetic field

KEK-SKS

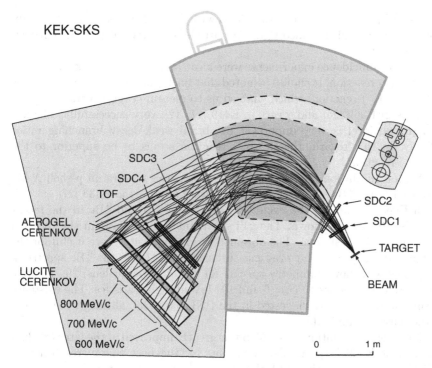

SDC3

SDC4

TOF

AEROGEL
CERENKOV

LUCITE
CERENKOV

800 MeV/c

700 MeV/c

600 MeV/c

SDC2

SDC1

TARGET

BEAM

0   1 m

Fig. 11. Schematic view of the SKS spectrometer. Typical particle trajectories are shown in the figure

of 3 T was constructed. Momentum reconstruction in the SKS relies on the magnetic field maps measured to a high precision of $\Delta B/B \ll 10^{-3}$ over the whole tracking region.

For incident momentum analysis, the last part of the beam line is composed of a QQDQQ spectrometer system. The $\langle x|\theta \rangle$ parameter of the QQDQQ transport matrix, which relates the position at the exit x_{out} to the incident angle $\theta_{\text{in}} \equiv dx/dz$ as $x_{\text{out}} = \langle x|\theta \rangle \times \theta_{\text{in}}$, is minimized so as to avoid a multiple-scattering effect on the momentum resolution.

The basic reaction mechanism and feasibility of the (π^+, K^+) reaction was theoretically investigated in 1980 [21]. The usefulness of the (π^+, K^+) reaction was first demonstrated at BNL-AGS in the $(\pi^+, K^+)^{12}_\Lambda$C reaction [22]. The measurement was further extended to various Λ hypernuclei as heavy as $^{89}_\Lambda$Y [23]. The excitation spectrum of the $(\pi^+, K^+)^{89}_\Lambda$Y reaction, in particular, beautifully showed the major shell structure of a Λ particle from the ground state to the f orbital with $\approx$3-MeV resolution, which was sometimes called a textbook example of the single-particle structure in nuclear physics [24].

The SKS has fully taken advantage of the usefulness of the (π^+, K^+) reaction with an improved energy resolution of $\leq$2 MeV. The first experiment on the (π^+, K^+) reaction was E140a, in which we obtained hypernuclear spectra of $^{10}_{\Lambda}$B, $^{12}_{\Lambda}$C, $^{28}_{\Lambda}$Si, $^{89}_{\Lambda}$Y, $^{139}_{\Lambda}$La, and $^{208}_{\Lambda}$Pb with 2–2.3 MeV$_{FWHM}$ energy resolution [25]. The result revealed that a Λ particle holds its single-particle nature even in $^{208}_{\Lambda}$Pb.

Several coincidence experiments were also carried out by using the SKS in the (π^+, K^+) reaction. It should be noted that background levels were quite low in the (π^+, K^+) reaction, which enabled us to measure even neutral particles, neutrons in E369 [26] and γ rays in E419 [11, 12], very successfully. The low background level is very important to obtain weak-decay branching ratios reliably. In this regard, the (π^+, K^+) reaction seems to be superior to the (K^-, π^-) reaction.

A good resolution (≈ 2 MeV) and high-statistics spectra on p-shell Λ hypernuclei ($^7_{\Lambda}$Li, $^9_{\Lambda}$Be, $^{12}_{\Lambda}$C, $^{13}_{\Lambda}$C, and $^{16}_{\Lambda}$O) were obtained in E336 [27].

In Figs. 12 and 13, the excitation spectra of $^7_{\Lambda}$Li and $^9_{\Lambda}$Be in the recent analysis [10, 28] are shown. The $^7_{\Lambda}$Li spectrum was obtained in the (π^+, K^+) reaction for the first time. The obtained production rates were useful for the design and analysis of the γ-ray measurement in E419. In the $^9_{\Lambda}$Be spectrum, the states with new symmetry specific for the Λ hypernucleus [29, 30] were observed as two peaks at $E_x = 6$ and 10 MeV in high statistics. The existence of such states was first suggested in the BNL data [23], although the quality of the data was limited.

The energy resolution of the SKS was greatly improved in the later (π^+, K^+) experiment E369 [31]. As shown in Fig. 14, the best energy resolution of 1.45 MeV$_{FWHM}$ was achieved for a 0.9-g/cm^2 carbon target.

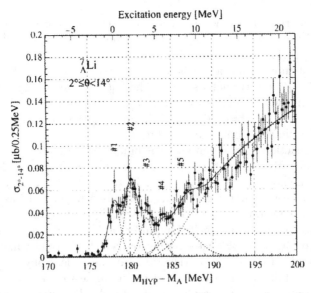

Fig. 12. Hypernuclear mass spectrum of $^7_{\Lambda}$Li obtained in E336 [10]

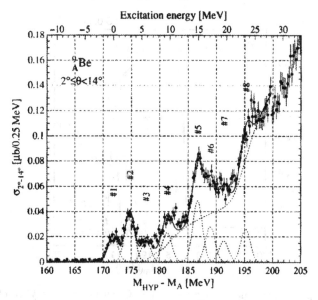

Fig. 13. Hypernuclear mass spectrum of $^9_\Lambda$Be obtained in E336 [10]

Even at an energy resolution of $1.9\,\mathrm{MeV_{FWHM}}$, we could clearly observe two core-excited states between the ground-state peak and the p_Λ peak at $E_x = 10.75\,\mathrm{MeV}$. The peak position of the second core-excited state was too high to interpret it as being the $[^{11}\mathrm{C}(3/2^-_2 ; 4.80\,\mathrm{MeV}) \otimes 1s^\Lambda](J=1^-_3)$ configuration, and the width was too large for a single peak. In the new spectrum, it was found that the peak is composed of two peaks: the first peak could be assigned to the $J=1^-_3$ state, and the second peak was ascribed to the state calculated by T. Motoba [32] in an extended model with intershell couplings. A small peak at $E_x = 12.4\,\mathrm{MeV}$, which was also first resolved in this spectrum, is ascribed to the $[^{11}\mathrm{C}(1/2^-_1 ; 2.00\,\mathrm{MeV}) \otimes 1p^\Lambda](J=2^+)$ configuration.

A high-statistics spectrum was also accumulated for the $(\pi^+, K^+)\,^{89}_\Lambda$Y reaction with an improved energy resolution of $1.65\,\mathrm{MeV}$. There had been no such high-quality data for heavy Λ hypernuclei before. As shown in Fig. 15, we could clearly find bump structures corresponding to the Λ major shell orbits (s, p, d, f) in the bound region. However, the widths of the bumps for the p, d, and f orbits were significantly wider than the experimental resolution. In fact, it was confirmed that the f orbit splits into two peaks.

In the figure, our result of the fitting is shown, in which we assumed the p, d, and f orbits were composed of two peaks (shown by dashed lines), and the s orbit was a single peak with the fixed width of the experimental resolution. Some contributions in between the major bumps were fitted with a series of Gaussian peaks with a wider width, taking account of the spreading width of such a deep neutron-hole state. The energy separations between two peaks in

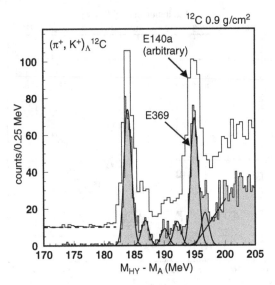

Fig. 14. Hypernuclear mass spectrum of $^{12}_{\Lambda}$C with an energy resolution of 1.45 MeV$_{\mathrm{FWHM}}$. An old spectrum with an energy resolution of 1.9 MeV is superimposed with an arbitrary scale and an offset for a comparison [31]

each orbit were obtained to be 1.37 ± 0.20, 1.63 ± 0.14, and 1.70 ± 0.10 MeV for the p, d, and f orbits, respectively. It should be noted that the separations become wider as the angular momenta of the Λ orbits increase.

A similar structure has been observed in the (π^+, K^+) $^{51}_{\Lambda}$V spectrum with an energy resolution of 1.95 MeV. In this case, the Λ orbits up to the d orbit are bound. Again, the widths of the bumps for the p orbit and the d orbit are broader than the experimental resolution.

So far, not many experimental efforts have been made to search for neutron-rich hypernuclei. When such a neutron-rich hypernucleus is formed, a Λ hyperon plays a glue-like role, and attracts the neutron-halo component to stabilize it. By using the SKS spectrometer, the first attempt to produce a $^{10}_{\Lambda}$Li hypernucleus was proposed by T. Fukuda as KEK-PS E521. A measurement of the ^{10}B(π^-, K^+) reaction at 1.05 GeV/c and 1.2 GeV/c was carried out successfully [33]. Although we could not identify any peaks for the bound states, the production of $^{10}_{\Lambda}$Li bound states has been clearly identified. The production rate was very low; compared to the normal (π^+, K^+) rate it was on the order of 10^{-3}. The production rate at 1.2 GeV/c was larger than that at 1.05 GeV/c. This suggests that production takes place through intermediate Σ states rather than two-step processes [34].

The spectroscopy of (π^+, K^+) has been successful with an energy resolution of 1.5–2 MeV, achieved with the SKS. It demonstrated that (π^+, K^+) spectroscopy is a unique way to investigate deeply-bound Λ states in heavy Λ hypernuclei. We expect many bound states with high angular momentum to

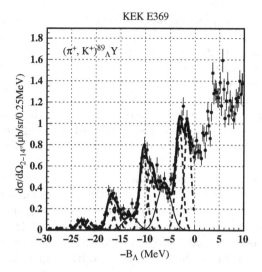

Fig. 15. Hypernuclear mass spectrum of $^{89}_{\Lambda}Y$ obtained in E369 with an energy resolution of 1.65 MeV$_{FWHM}$. It was found that the peak for the f_{Λ} orbit splits into two peaks [31]

appear in a deep potential of $U_{\Lambda} \approx 30$ MeV in heavy Λ hypernuclei. Measuring the excitation energy with high precision would be interesting to examine the single-particle nature of a Λ particle deeply bound in a heavy nucleus. A high energy resolution of ≤ 1 MeV would be needed to resolve most of the Λ single-particle orbits.

5 Examples of Hypernuclear Spectroscopy

5.1 Hypernuclear γ-ray Spectroscopy

BNL-AGS E929

A unique measurement of E1 γ-ray transition in $^{13}_{\Lambda}C$ was carried out in E929 at BNL-AGS [35]; the splitting between $p_{1/2,\Lambda}$ and $p_{3/2,\Lambda}$ orbits was obtained by measuring the two γ-ray energies corresponding to the inter-shell transitions of Λ, $p_{1/2,\Lambda} \rightarrow s_{1/2,\Lambda}$ and $p_{3/2,\Lambda} \rightarrow s_{1/2,\Lambda}$.

In $^{13}_{\Lambda}C$, the two p orbitals, $1/2^-$ and $3/2^-$ are just below the Λ emission threshold, and there are no strong decays below these levels. Therefore, the γ rays can be observed. While in the case of $^{12}_{\Lambda}C$, a 2^- p_{Λ} state decays into $^{11}_{\Lambda}B+p$ and no such γ-ray transitions are observed.

One problem here is how to produce both p states. As I explained in Sect. 2.1, the (K^-, π^-) reaction at forward angles excites the $p_{1/2,\Lambda}$ state dominantly in $^{13}_{\Lambda}C$, while the $p_{3/2,\Lambda}$ state very weakly. The so-called substitutional transition, $p_{1/2,n} \to p_{1/2,\Lambda}$, is dominant in such a low momentum transfer reaction. A solution is to excite the $p_{3/2,\Lambda}$ state with transfer of two units of angular momentum ($\Delta\ell=2$), which is expected to dominate in the π^-'s scattering angles of 10° to 20°.

The measurement was carried out with the $^{13}C\,(K^-,\pi^-)\,^{13}_{\Lambda}C$ reaction at 0.93 GeV/c at the D6 beam line of BNL-AGS. A large solid angle magnet (48D48) was used to measure the π^- momentum in the large angular range, $\approx 0°$–$16°$. γ rays for the $^{13}_{\Lambda}C$ were measured with two NaI detector arrays; each array consisted of 36 NaI crystals, each of which had a dimension of 6.5 $\times$ 6.5 $\times$ 30 cm^3.

The energy spectra of γ rays in coincidence with the scattered π^-'s at $0° \leq \theta_\pi \leq 7°$, $7° \leq \theta_\pi \leq 10°$, and $10° \leq \theta_\pi \leq 16°$ are shown in Fig. 16 (upper panel). The peaks at ≈ 11 MeV correspond to the p_Λ-to-s_Λ transitions. Here, it is assumed that two γ rays are mixed in these peaks. The mixing ratios were estimated based on a distorted-wave impulse approximation shown in Fig. 16 (lower panel). From the γ ray peak positions and the mixing ratios in three scattering angles, the energy difference between $p_{1/2,\Lambda} \to s_{1/2,\Lambda}$ and $p_{3/2,\Lambda} \to s_{1/2,\Lambda}$ transitions was obtained to be $152 \pm 54\,(\text{stat}) \pm 36\,(\text{syst})$ keV.

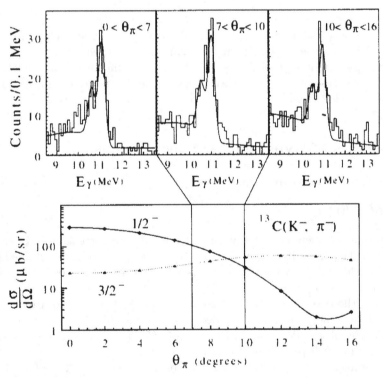

Fig. 16. γ-ray spectra in coincidence with scattered π^-'s (*upper panel*) and theoretical curves of differential cross section for $1/2^-$ and $3/2^-$ states (*lower panel*) are shown [35]

The result indicates the spin-orbit splitting in $^{13}_{\Lambda}$C is very small compared with the typical 3-5 MeV splitting in ordinary nuclei around this mass region.

KEK-PS E419

In KEK-PS E419 experiment, hypernuclear γ rays from $^{7}_{\Lambda}$Li were measured in high resolution by using a large-acceptance germanium (Ge) detector array called Hyperball. The $^{7}_{\Lambda}$Li was produced via the (π^+, K^+) reaction on a ^{7}Li target (25 cm thick, 98% enriched). In the (π^+, K^+) reaction, it was already known [36] that the four states $(1/2^+(T = 0), 5/2^+(T = 0), 1/2^+(T = 1)$, and $5/2^+(T = 1))$ of $^{7}_{\Lambda}$Li are strongly populated (see Fig. 17). In fact, four hypernuclear γ-ray transitions shown in Fig. 17 were observed, while two additional γ rays at 429 keV and 478 keV were observed at the same time [37]. They were interpreted as the γ transitions in the daughter nuclei resulting from the $^{7}_{\Lambda}$Li weak decays: $^{7}_{\Lambda}$Li $\rightarrow {}^7$Be*$(1/2^-, 429\,\text{keV}) + \pi^-$ and $^{7}_{\Lambda}$Li $\rightarrow {}^7$Li*$(1/2^-, 478\,\text{keV}) + \pi^0$ as shown in the decay scheme in Fig. 17.

The ground-state spin of the $^{7}_{\Lambda}$Li is sensitive to the mesonic weak decay rate, because the π-mesonic weak decay of Λ takes place dominantly with the s-wave (spin-non-flip) amplitude compared with the p-wave (spin-flip) amplitude as explained in Sect. 2.2. The decay rates of $^{7}_{\Lambda}$Li to the first excited state of ^{7}Be were calculated in [38] for two cases of the $^{7}_{\Lambda}$Li spin of $1/2^+$ and $3/2^+$ as,

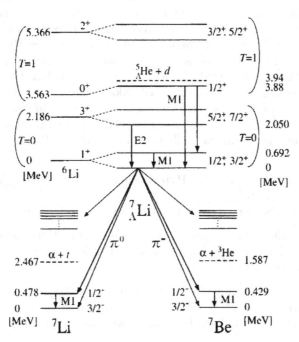

Fig. 17. Level scheme of γ transitions and mesonic weak decays of $^{7}_{\Lambda}$Li [37]

$$\Gamma(_\Lambda^7\text{Li}(1/2^+) \rightarrow {}^7\text{Be}^* + \pi^-) = 0.070\,\Gamma_\Lambda\,,$$

$$\Gamma(_\Lambda^7\text{Li}(3/2^+) \rightarrow {}^7\text{Be}^* + \pi^-) = 0.007\,\Gamma_\Lambda\,,$$

where Γ_Λ denotes the total decay rate of a free Λ. The $1/2^+$ and $3/2^+$ members of the ground-state doublet of $_\Lambda^7\text{Li}$ have $L = 0$ with $S = 1/2$ and $S = 3/2$ configuration, mainly, while the first excited state ${}^7\text{Be}^*(1/2^-)$ has $L = 1$ and $S = 1/2$. Thus, the spin-non-flip transition of $_\Lambda^7\text{Li}(1/2^+)$ to ${}^7\text{Be}^*(1/2^-) + \pi^-$ is preferred compared with the spin-flip transition from the $_\Lambda^7\text{Li}(3/2^+)$.

Assuming the total decay rate to be $\Gamma_\text{tot}(_\Lambda^7\text{Li}) \approx (1.2 \pm 0.4)\,\Gamma_\Lambda$, the theoretical branching ratios are estimated to be $(5.8 \pm 1.9) \times 10^{-2}$ for the $1/2^+$ case, and $(0.6 \pm 0.2) \times 10^{-2}$ for the $3/2^+$ case. The measured value was $(6.0^{+1.3}_{-1.6}) \times 10^{-2}$. It agrees very well with that for the $1/2^+$ case, and disagrees with that for the $3/2^+$ case. Therefore, the ground-state spin of $_\Lambda^7\text{Li}$ was determined to be $1/2$.

6 New Hypernuclear Spectroscopy at J-PARC

A new high-intensity proton accelerator facility called J-PARC has been in construction in Japan since 2001. The accelerator consists of a proton linac, a rapid-cycling (25 Hz) 3-GeV proton synchrotron, and a 50-GeV proton synchrotron (main ring). The proton beam from the main ring, 30 GeV with 9 μA in the initial stage, will produce various beams of kaons, pions, neutrinos, and antiprotons of high intensity. A neutrino beam line for fast extraction and an experimental area for slow extraction, called Hadron Experimental Area, will be constructed for Nuclear and Particle physics experiments at the main ring.

In November 2005, the proposals for Nuclear and Particle physics experiments at J-PARC were called. Twenty proposals including four letters of intent were received by the end of April, 2006. Around 10 were related to Strangeness Nuclear Physics. The 14 proposals were considered at the first PAC meeting held at the end of June. The PAC approved three experiments at this time: one for neutrino oscillations experiment called T2K, and the other two are:

E05: Spectroscopic Study of Ξ Hypernucleus, $_\Xi^{12}\text{Be}$, via the ${}^{12}\text{C}\,(\text{K}^-, \text{K}^+)$
    Reaction (T. Nagae[1]),
E13: Gamma-ray Spectroscopy of Light Hypernuclei (H. Tamura).

As well as these two approved experiments, the Committee also selected the following five experiments as stage-1 (scientific merit) approval in Strangeness Nuclear Physics;

E03: Measurement of X Rays from Ξ^- Atom (K. Tanida),
E07: Systematic Study of Double Strangeness System with an Emulsion-counter Hybrid Method (K. Imai, K. Nakazawa, H. Tamura),

---

[1] The name(s) in parentheses is (are) Spokesperson(s).

E15: A Search for Deeply-bound Kaonic Nuclear States by In-flight ^{3}He(K$^-$, n)
    Reaction (M. Iwasaki, T. Nagae),

E17: Precision Spectroscopy of Kaonic ^{3}He 3d→2p X-rays (R.S. Hayano,
    H. Outa),

E19: High-resolution Search for Θ^+ Pentaquark in π^-p → K$^-$X Reaction
    (M. Naruki).

The experiments E05, E13, E15, E17, and E19 were also categorized as Day-1
experiments in Hadron Experimental Area. Among them, E05 has the first
priority and E13 the second priority.

Here, I would like to introduce three experiments, E05, E13, and E15.

6.1 E05: Spectroscopic Study of Ξ Hypernuclei

The high intensity K$^-$ beam at ≈ 1.8 GeV/c available at J-PARC Hadron Facility is quite unique to open a new frontier of strangeness nuclear physics in
the spectroscopic studies of strangeness $S = -2$ systems; here, the $S = -2$ systems include Ξ hypernuclei, double-Λ hypernuclei, and possibly H hypernuclei.
This is not only a step forward from $S = -1$ systems as a natural extension,
but also a significant step to explore the multi-strangeness hadronic systems;
in the course of the limit, strange hadronic matter ($S = -\infty$) in the core of a
neutron star is our concern. It is also important to extract quantitative information on ΞN and ΛΛ interactions from the spectroscopic data, considering
the fact that there exists almost no data on these interactions at this moment.
Hence, we can explore the SU(3)$_f$ character of the strong forces of QCD.

Previous Studies on $S = -2$ Systems

The (K$^-$, K$^+$) reaction is one of the best tools to implant the $S = -2$ through
the elementary process, K$^-$p → K$^+$Ξ$^-$, the cross section of which in the
forward angle has a broad maximum around the momentum of 1.8 GeV/c [39].
This reaction has been used for studies of $S = -2$ systems so far.

    As for the Ξ hypernuclei, there exist some hints of emulsion events for
the existence. However it is still not conclusive. Some upper limits on the Ξ-
nucleus potential have been obtained from the production rate and spectrum
shape in the bound region of Ξ hypernucleus via ^{12}C(K$^-$, K$^+$) reaction [40, 41].
In these experiments, Ξ-hypernuclear states were not clearly observed because
of the limited statistics and detector resolution (Fig. 18). As shown in the right
figure, the potential depth, V_Ξ, is favored to be ≈ -14 MeV for A = 12 when
a Woods-Saxon type potential shape is assumed.

    As for double-Λ hypernuclei, several emulsion events were reported [42,
43, 44, 45, 46]. Among them, however, Nagara event recently found in a
hybrid-emulsion experiment, KEK-PS E373, was able to cleanly identify the
$_{\Lambda\Lambda}^6$He [46]. The mass of the $_{\Lambda\Lambda}^6$He and thus Λ-Λ interaction energy, $\Delta B_{\Lambda\Lambda}$,
has been measured for the first time. It demonstrated that Λ-Λ interaction

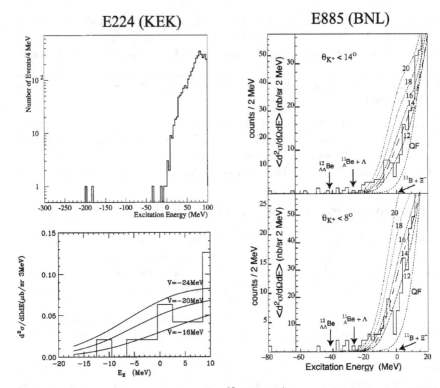

Fig. 18. The missing mass spectra for ^{12}C(K$^-$, K$^+$) reaction obtained in KEK-E224 (*left*) and BNL-E885 (*right*) taken from [40] and [41]. Curves in the figures are calculated spectra using various potential depths taking the experimental resolutions into account

is weakly attractive; weaker than estimated before. The production of $_{\Lambda\Lambda}^{4}$H was also reported in a counter experiment by detection of pairs of pions in sequential mesonic weak decays [47]. However, $\Delta B_{\Lambda\Lambda}$ was not well determined due to poor statistics and the insufficient resolution.

Spectroscopy of Ξ Hypernuclei and $S=-2$ Systems

The Ξ hypernuclei play an important role in the investigations of $S=-2$ systems as the entrance channel to the $S=-2$ world. In Fig. 19, typical energy spectrum and decay thresholds for Ξ- and double-Λ hypernuclear configurations are shown. Produced Ξ-hypernuclear states eventually decay into several forms of double-Λ systems through a strong conversion process, $\Xi^-p \to \Lambda\Lambda$. Moreover, Ξ-hypernuclei give valuable information on the $S=-2$ baryon-baryon interactions such as ΞN, and $\Xi N \to \Lambda\Lambda$. At this moment, we still even don't know whether the ΞN interaction is attractive or not, and thus,

Energy Spectrum of S=-2 systems

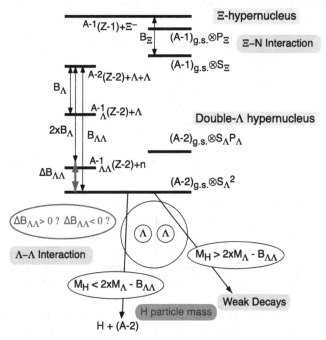

Fig. 19. Typical energy spectrum and decay thresholds for $S = -2$ system; Ξ and double-Λ hypernuclear configurations [48]

whether Ξ hypernuclei really exist or not. Although, the BNL E885 claims the evidence [41].

While single-Λ and double-Λ hypernuclear ground states decay via weak interaction and therefore they are long-lived, Ξ-hypernuclei decay via the strong interaction through a conversion process $\Xi^- p \to \Lambda\Lambda$ ($Q = 28.3$ MeV). This situation is very similar to Σ hypernuclei in which the strong conversion process $\Sigma N \to \Lambda N$ ($Q \approx 75$ MeV) also exists and broadens the width of the states. However, $\Xi^- p \to \Lambda\Lambda$ conversion occurs in the $^1S_0(T=0)$ state, a weight of which is only $1/16$ in nuclear matter. Although it depends on the interaction models, the width for finite nuclei may be reduced to ≤ 1 MeV due to the reduction of the phase space and the reduction of an overlap of the wave functions. The calculated widths in several interaction models are listed in Table 1 for nuclear matter.

Therefore, it is expected that the spectroscopy of the Ξ hypernuclei is promising. Here we use the (K^-, K^+) reaction in which we can use the same method as in the (π^+, K^+) reaction successful for the Λ-hypernuclear spectroscopy. A Ξ^- is produced with a large recoil momentum similar to a Λ for the (π^+, K^+) reaction; $p_\Xi \approx 500$ MeV/c (while $p_\Lambda \approx 350$ MeV/c). Due

Table 1. Ξ potentials U_Ξ and partial wave contributions in nuclear matter at normal density, calculated with NHC-D, Ehime and ESC04d*. Conversion widths in nuclear matter, Γ_Ξ, is also listed (in MeV)

Model	T	$^1S_0$	$^3S_1$	$^1P_1$	$^3P_0$	$^3P_1$	$^3P_2$	U_Ξ	Γ_Ξ
NHC-D	0	−2.6	0.1	−2.1	−0.2	−0.7	−1.9		
	1	−3.2	−2.3	−3.0	−0.0	−3.1	−6.3	−25.2	0.9
Ehime	0	−0.9	−0.5	−1.0	0.3	−2.4	−0.7		
	1	−1.3	−8.6	−0.8	−0.4	−1.7	−4.2	−22.3	0.5
ESC04d*	0	6.3	−18.4	1.2	1.5	−1.3	−1.9		
	1	7.2	−1.7	−0.8	−0.5	−1.2	−2.8	−12.1	12.7

to the large momentum transfer of the reaction, spin-stretched states can be selectively populated as in the case of Λ hypernuclei by the (π^+, K^+) reaction. This selectivity helps us to observe well-separated peak structures among many possible excitations even for heavy targets. The cross section of the elementary $(K^-, K^+)\,\Xi^-$ process is considerably smaller than that of the $n(\pi^+, K^+)\Lambda$ reaction; $35\,\mu b/sr$ for the (K^-, K^+) , while $\approx 500\,\mu b/sr$ for the (π^+, K^+).

In spite of several demerits, the (K^-, K^+) reaction is the best tool for the reaction spectroscopy of Ξ hypernuclei. The previous studies were not able to clearly conclude the existence of the Ξ-hypernuclear states. This is because their detector resolution was not sufficient to identify the states and the statistics was not enough to detect the states with such small cross sections.

Therefore it is desirable to perform the reaction spectroscopy again by using the high-intensity K^- beams at J-PARC and the spectrometers with a much improved energy resolution.

Since little is known for $S = -2$ baryon-baryon systems, especially the ΞN system, there is no established interaction model for the $S = -2$ channels. This is mainly due to the lack of experimental data such as ΞN elastic and inelastic scattering. As a consequence, the derived one-body Ξ potentials, U_Ξ, are remarkably different among the available interaction models. In some cases, they are even repulsive.

Here, the calculation results [49] for three different interaction models are presented:

- Nijmegen Hard-Core model D (NHC-D) [50]
- Ehime model [51]
- Extended Soft-Core model 04d* (ESC04d*) [52]

These models give an attractive (negative) value of U_Ξ.

Table 1 shows the calculated values of the Ξ potential (U_Ξ), conversion width (Γ_Ξ) and partial wave contribution in nuclear matter obtained in the G-matrix calculations with three interaction models.

The NHC-D and Ehime models predict deep U_Ξ and strong mass-number (A) dependence for Ξ energies. This is owing to the strong odd-state attractions which come from the lack of space-exchange terms in one-boson-exchange potential picture. On the other hand, the ESC04d* model predicts the energies very close to those with the Woods-Saxon potential with a well depth of $\approx 14\,\text{MeV}$. Therefore, the experimental data on not only the Ξ-binding energies but also their A-dependence contains valuable information to probe the ΞN interaction. It should be also noted that the conversion widths for the ESC04d* and NHC-D/Ehime models are very different from each other.

Knowledge of the depth of the Ξ-nucleus potential is also important for estimating the existence of strange hadronic matter with Ξ [53]. For a long time, it was believed that Σ^- hyperons would appear in neutron stars earlier (i.e. at lower density) than even lighter Λ hyperons due to their negative charge. However, recent data [54] strongly suggest that the interaction of the Σ^- with neutron-rich nuclear systems is strongly repulsive, which means that Σ^- hyperons can no longer appear in neutron stars. It was argued that the disappearance of Σ^- does not necessarily lead to crucial changes of neutron star features if they were substituted effectively by Ξ^- hyperons. However better understanding of the ΞN interaction is necessary for a definite conclusion. With respect to the structure of the neutron star, it becomes much more important to investigate the Ξ dynamics than it was considered before the Σ-nucleus repulsion has been established.

Experimental Methods and Apparatus

The main physics goal of the experiment is to obtain the conclusive results on the existence of Ξ hypernuclei by observing bound states of $^{12}_{\Xi}\text{Be}$ hypernucleus via the $^{12}\text{C}(K^-, K^+)$ reaction with the best energy resolution of a few MeV so far achieved and in good statistics. For this purpose, we need two high-resolution spectrometers for K^- beam and scattered K^+'s.

In the previous measurement, the potential depth of $\approx 14\,\text{MeV}$ was extracted from the shape analysis near the Ξ binding threshold with a help of theoretical calculations. This estimated potential depth suggests that we could definitely observe the bound state peak distinguished from the quasifree continuum as shown in Fig. 20. It demonstrates the importance of the good resolution of the measurements when one compares it with the right-top of Fig. 18. A peak position will give us more direct information on the depth of the Ξ-nucleus potential than the previous shape analysis. The width of the bound state peak also provides us with information on the imaginary part of the Ξ-nucleus potential, or the ΞN inelastic channel.

A new kaon beam line K1.8 with the maximum beam momentum of $1.8\,\text{GeV}/c$ has been designed for the experiment. The beamline has two characteristic features; (1) high-intensity ($1.4 \times 10^6\,\text{K}^-/\text{spill}$) and high-purity K^- beams are obtained. (2) a high resolution beam analyzer is located at the end of the beamline. These features are required and optimized to perform the

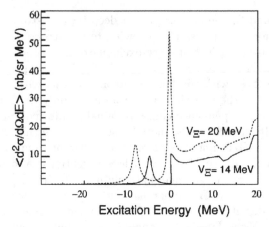

Fig. 20. The calculated spectrum for ^{12}C (K$^-$, K$^+$) $^{12}_\Xi$Be reaction for Ξ potential depths of 14 MeV (*solid*) and 20 MeV (*dotted*) taken from [41]. The cross section has been averaged over the kaon angular range from $0°$ to $14°$

spectroscopic studies on Ξ hypernuclei. The beamline has two stages of electrostatic separators with two mass slits in order to separate kaons from pions and other particles at the level of K$^-$/π^- ratio greater than 5. The beam analyzer located after the last mass slit comprises QQDQQ magnets and four sets of tracking detectors. The expected momentum resolution $\Delta p/p$ is 1.4×10^{-4} in root-mean-square when a position resolution of $200\,\mu$m is realized in the tracking detectors placed before and after the QQDQQ system.

For the K$^+$ spectrometer, the existing SKS spectrometer will be used with some modifications. The K$^+$ momentum corresponding to the production of Ξ hypernuclei is around 1.3 GeV/c. The SKS maximum magnetic field of ≈ 2.7 T does not allow us to put the central ray at 1.3 GeV/c. Therefore, the central ray is shifted outer side and a dipole magnet with ≈ 1.5 T will be added at the entrance of the SKS magnet as shown in Fig. 21. A simulation shows that the spectrometer, (called SKS+), has a solid angle of ≈ 30 msr with the angular range up to $10°$, and momentum resolution $\Delta p/p_{\mathrm{FWHM}} = 0.17\%$.

The overall energy resolution is expected to be better than $3\,\mathrm{MeV_{FWHM}}$ including the energy-loss straggling in the target.

The production cross sections of the Ξ hypernuclei in the (K$^-$, K$^+$) reaction have been calculated by several theorists within the framework of the distorted-wave impulse approximation (DWIA) [48, 55, 56]. Also, the previous experimental studies reported the cross sections [40, 41]. Based on these, the yield of $^{12}_\Xi$Be is estimated to be ≈ 190 events/month.

6.2 E13: Gamma-ray Spectroscopy of Light Hypernuclei

A lot of γ-ray transitions for various Λ hypernuclei are expected to be observed at J-PARC. Abundant productions of Λ hypernuclei could be possible in the (K$^-$, π^-) reactions by using high intensity K$^-$ beams.

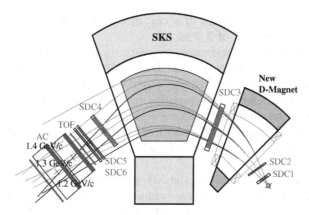

Fig. 21. SKS+ Spectrometer in consideration

As for the Day 1 experiment, the (K^-, π^-) reaction at $1.5\,\text{GeV}/c$ will be used to produce Λ hypernuclei at the K1.8 beam line together with the SKS spectrometer. The incident beam momentum was selected considering the available beam intensity at the K1.8 beam line, momentum dependence of the production cross sections, and the spin-flip amplitude of the (K^-, π^-) reaction. The maximum K^- beam intensity at Day 1 would be 0.5×10^6 per spill at $1.5\,\text{GeV}/c$ at the K1.8 beam line. The SKS magnet should be operated at $2.7\,\text{T}$ for the scattered pion momentum of $\approx 1.4\,\text{GeV}/c$. The tracking detectors at the exit of the SKS magnet should be replaced with larger ones to keep the solid angle acceptance $\geq 100\,\text{msr}$. The hypernuclear mass resolution is expected to be around $6\,\text{MeV}$ which is mainly due to energy-loss straggling in a target.

A new germanium detector Hyperball-J is going to be constructed for the experiment. It consists of about thirty sets of Ge detectors having a relative photo-peak efficiency of about 75%. Each Ge detector is surrounded with fast PWO counters instead of the previous BGO counters in Hyperball, for background suppression. The photo-peak efficiency is expected to be better than 5% at $1\,\text{MeV}$ at the distance of $\approx 15\,\text{cm}$ from a target. Detailed design work is now in progress.

Various interesting subjects are proposed by using the Hyperball-J.

One of the important subjects is to measure the transition probabilities $(B(M1))$ of the Λ spin-flip M1 transitions, and probe the g factor of a Λ inside a nucleus. We measure the $M1(3/2^+ \rightarrow 1/2^+)$ transition of $^7_\Lambda\text{Li}$, where the $3/2^+$ state is populated from the $1/2^+(T=1)$ state via the fast $1/2^+(T=1) \rightarrow 3/2^+$ transition. The $1/2^+(T=1)$ state is populated by the (K^-, π^-) non-spin-flip reaction with a large cross section.

The lifetime of the $3/2^+$ state is estimated to be $\approx 0.5\,\text{ps}$. In order to apply Doppler-Shift Attenuation Method [12] to measure the lifetime, the stopping time of the recoiling $^7_\Lambda\text{Li}^*$ should be 2–3 times longer than the lifetime. Thus, a Li_2O target with a density of $2.01\,g/cm^3$ in granular powder is selected with a calculated stopping time of 2–3 ps.

In the studies of the Λ hypernuclear structure and ΛN interaction, there are still puzzles to be solved even in the p-shell region. While γ-ray measurements were performed on $^{10}_{\Lambda}B$ and $^{11}_{\Lambda}B$, the observations were hardly understood. Much detailed studies with high statistics and $\gamma-\gamma$ coincidence measurements would be needed to construct the level schemes. Also, it is proposed to take the data with a ^{19}F target to detect both $^{19}_{\Lambda}F(1/2^- \rightarrow 3/2^+, 1/2^+)$ transitions. It would determine the ground-state doublet spacing in the sd-shell region for the first time.

It is also planned to measure the M1 γ ray in $^{4}_{\Lambda}He(1^+ \rightarrow 0^+)$ with high precision. An extremely large charge-symmetry breaking is reported between $^{4}_{\Lambda}He$ and $^{4}_{\Lambda}H$. An improved measurement would clarify the effect. Also, this measurement is useful to confirm the spin-flip and non-spin-flip amplitudes in the (K^-, π^-) reaction in this momentum range.

These series of γ-ray measurements could be carried out in a few months of data taking at J-PARC.

6.3 E15: A Search for Deeply-Bound Kaonic Nuclear States

The existence of deeply-bound kaonic nuclear states, which were suggested as narrow states for specific finite nuclear systems [57], is now an important experimental subject to be confirmed. There exist several experimental data suggesting the existence of such bound states [58, 59]. However, from the theoretical side, there has been a lot of discussions whether the K^- nucleus potential is deeply attractive ($-\mathrm{Re}V_{opt}(\rho_0) \approx 150\text{--}200\,\mathrm{MeV}$) [60, 61] or much shallower ($-\mathrm{Re}V_{opt}(\rho_0) \approx 50\text{--}75\,\mathrm{MeV}$) [62, 63, 64, 65, 66]. Both types of potentials reasonably reproduce the shifts and widths of the kaonic X-ray data [67]. There are also discussions that the widths of the bound states would be too broad to be separately observed as a clean bound-state peak. Moreover, there is an issue whether such a deeply bound state, if existed, gives rise to a formation of a dense nuclear system by strongly attracting the system. If this is true, the formation of deeply bound states would give us a unique opportunity to investigate a hadron in dense nuclei.

Therefore, it is an urgent task to confirm whether such a bound state exists or not, experimentally.

In this experiment, we are going to measure the mass of K^-pp system both in the missing-mass measurement and in the invariant-mass measurement. The K^-pp system is important because it would be the simplest and lightest kaon bound state, if existed. Therefore, a clean and unambiguous identification of the bound state could be possible. Here, we use the in-flight (K^-, n) reaction on ^3He at $1\,GeV/c$, where the cross section of $K^- + n \rightarrow n + K^-$ has a broad maximum of $\approx 5\,mb/sr$. The K^-pp mass is measured as a missing-mass expressed as

$$M_{\mathrm{K^-pp}} = \sqrt{(E_{\mathrm{n}} - E_{\mathrm{K^-}} - M_{^3\mathrm{He}})^2 - (\boldsymbol{p}_{\mathrm{n}} - \boldsymbol{p}_{\mathrm{K^-}})^2} \,. \tag{7}$$

At the same time, the target region is covered with a cylindrical detector system with a large acceptance and a solenoidal magnetic field. Thus, most of the charged particles produced in the decay of the $\mathrm{K^-pp}$ system are detected. Here, a decay mode of $\mathrm{K^-pp} \rightarrow \Lambda + \mathrm{p}$ followed by the $\Lambda \rightarrow \mathrm{p} + \pi^-$ decay is detected, and the $\mathrm{K^-pp}$ mass is reconstructed as an invariant mass of a Λ and a proton.

A designed missing-mass resolution is $\approx 28\,(\mathrm{MeV}/c^2)_{\mathrm{FWHM}}$ with a flight path of $\approx 12\,\mathrm{m}$, and the invariant mass resolution is $\approx 40\,(\mathrm{MeV}/c^2)_{\mathrm{FWHM}}$.

7 Summary

A lot of experimental efforts to improve the mass resolution of hypernuclear spectroscopy are in progress all over the world. In JLab, high resolution magnetic spectrometers for $(\mathrm{e}, \mathrm{e'\,K^+})$ reactions were constructed with 0.7–0.9 MeV resolutions. And, a new spectrometer aiming at $\approx 0.3\,\mathrm{MeV}$ resolution is now under construction. In FINUDA experiments at Frascati, the energy resolution of 1.3 MeV is achieved for the $(\mathrm{K^-_{stop}}, \pi^-)$ reactions.

The success of hypernuclear γ-ray spectroscopy with Hyperball is very remarkable with a few keV resolution. Systematic spectroscopic investigations could be carried out at J-PARC by using the high-intensity $\mathrm{K^-}$ beams in the near future. γ-coincidence measurements will be very useful for precise hypernuclear spectroscopy.

Further, new spectroscopic studies on $S = -2$ systems will be conducted at J-PARC with the $(\mathrm{K^-}, \mathrm{K^+})$ reactions: Ξ hypernuclei and double-Λ hypernuclei. These would be the first step toward the study of multi-strangeness systems.

References

1. T. Nagae: *17th Int. Conf. on Particles and Nuclei, PANIC05*, Santa Fe, NM, Conference Proceedings No. 842, ed by P. D. Barnes, M. D. Cooper, R. A. Eisenstein, H. van Hacke, and G. J. Stephenson (American Institute of Physics, New York, 2005) pp. 1021–1029
2. T. Motoba, H. Bandō, T. Fukuda, and J. Žofka: Nucl. Phys. A **534**, 597 (1991)
3. H. Bandō, T. Motoba, and J. Žofka: Int. J. Mod. Phys. A **5**, 4021 (1990)
4. R. Bertini et al: Nucl. Phys. A **360**, 315 (1981)
5. D. H. Davis: Nucl. Phys. A **754**, 3c (2005)
6. M. M. Block, L. Lendinara, and L. Monari: Proc. *Int. Conf. on High-Energy Physics*, CERN (1962) pp. 371–372
7. R. H. Dalitz and L. Liu: Phys. Rev. **116**, 1312 (1959)
8. D. Bertrand et al: Nucl. Phys. B **16**, 77 (1970)
9. W. Brückner et al: Phys. Lett. **55B**, 107 (1975); **62B**, 481 (1976); **79B**, 157 (1978)

10. O. Hashimoto and H. Tamura: Prog. Part. Nucl. Phys. **57**, 564 (2006)
11. H. Tamura et al: Phys. Rev. Lett. **84**, 5963 (2000)
12. K. Tanida et al: Phys. Rev. Lett. **86**, 1982 (2001)
13. A. Yamamoto: *Study of low energy intense kaon beam*, KEK Report KEK81-13 (1981)
14. P. H. Pile et al: Nucl. Instrum. Methods Phys. Res. A **321**, 48 (1992)
15. M. Danysz and J. Pniewski: Phil. Mag. **44**, 348 (1953)
16. R. E. Chrien: Nucl. Phys. A **691**, 501c (2001)
17. E. V. Hungerford: Nucl. Phys. A **691**, 21c (2001)
18. M. Agnello et al: Phys. Lett. B **622**, 35 (2005)
19. T. Fukuda et al: Nucl. Instrum. Methods Phys. Res. A **361**, 485 (1995)
20. T. Nagae: Nucl. Phys. A **691**, 76c (2001)
21. C. B. Dover, L. Ludeking, and G. E. Walker: Phys. Rev. C **22**, 2073 (1980)
22. C. Milner et al: Phys. Rev. Lett. **54**, 1237 (1985)
23. P. H. Pile et al: Phys. Rev. Lett. **66**, 2585 (1991)
24. D. J. Millener, C. Dover, and A. Gal: Phys. Rev. C **38**, 2700 (1988)
25. T. Hasegawa et al: Phys. Rev. C **53**, 1210 (1996)
26. J. H. Kim et al: Phys. Rev. C **68**, 065201 (2003)
27. S. Ajimura et al: Nucl. Phys. A **639**, 93c (1998)
28. H. Noumi: Nucl Phys. A **691**, 123c (2001)
29. T. Motoba, H. Bandō, and K. Ikeda: Prog. Theor. Phys. **70**, 189 (1983)
30. R. H. Dalitz and A. Gal: Phys. Rev. Lett. **36**, 362 (1976)
31. H. Hotchi et al: Phys. Rev. C **64**, 044302 (2001)
32. T. Motoba: Nucl. Phys. A **639**, 135c (1998)
33. P. K. Saha et al: Phys. Rev. Lett. **94**, 052502 (2005)
34. T. Yu. Tretyakova and D. E. Lanskoy: Nucl. Phys. A **691**, 51c (2001)
35. S. Ajimura et al: Phys. Rev. Lett. **86**, 4255 (2001);
    H. Kohri et al: Phys. Rev. C **65**, 034607 (2002)
36. R. H. Dalitz and A. Gal: Ann. Phys. (NY) **116**, 167 (1978)
37. J. Sasao et al: Phys. Lett. B **579**, 258 (2004)
38. T. Motoba, K. Itonaga, and H. Bandō: Nucl. Phys. A **489**, 683 (1988);
    T. Motoba and K. Itonaga: Prog. Theor. Phys. Suppl. **117**, 477 (1994)
39. C. B. Dover and A. Gal: Ann. of Phys. **146**, 309 (1983)
40. T. Fukuda et al: Phys. Rev. C **58**, 1306 (1998)
41. P. Khaustov et al: Phys. Rev. C **61**, 054603 (2000)
42. M. Danysz et al: Nucl. Phys. **49**, 121 (1963)
43. R. H. Dalitz et al: Proc. Roy. Soc. Lond. **A426**, 1 (1989)
44. D. J. Prowse: Phys. Rev. Lett. **17**, 782 (1966)
45. S. Aoki et al: Prog. Theor. Phys. **85**, 1287 (1991)
46. H. Takahashi et al: Phys. Rev. Lett. **87**, 212502 (2001)
47. J. K. Ahn et al: Phys. Rev. Lett. **87**, 132504 (2001)
48. C. B. Dover, A. Gal, and D. J. Millener: Nucl. Phys. A **572**, 85 (1994)
49. Y. Yamamoto: private communication
50. M. M. Nagels, Th. A. Rijken, and J. J. de Swart: Phys. Rev. D **15**, 2547 (1977)
51. M. Yamaguchi, K. Tominaga, Y. Yamamoto, and T. Ueda: Prog. Theor. Phys. **105**, 627 (2001)
52. Th. A. Rijken and Y. Yamamoto: Phys. Rev. C **73**, 044008 (2006)
53. J. Schaffner-Bielich and A. Gal, Phys. Rev. C **62**, 034311 (2000)
54. P. K. Saha et al: Phys. Rev. C **70**, 044613 (2004)

55. K. Ikeda, T. Fukuda, T. Motoba, M. Takahashi, and Y. Yamamoto: Prog. Theor. Phys. **91**, 747 (1994); Y. Yamamoto, T. Motoba, T. Fukuda, M. Takahashi, and K. Ikeda: Prog. Theor. Phys. Suppl. **117**, 281 (1994)
56. S. Tadokoro, H. Kobayashi, and Y. Akaishi: Phys. Rev. C **51**, 2656 (1995)
57. Y. Akaishi and T. Yamazaki: Phys. Rev. C **65**, 044005 (2002)
58. M. Agnello et al: Phys. Rev. Lett. **94**, 212303 (2005)
59. T. Kishimoto et al: Nucl. Phys. A **754**, 383c (2005)
60. E. Friedman, A. Gal, J. Mareš, and A. Cieplý: Phys. Rev. C **60**, 024314 (1999)
61. E. Friedman, A. Gal, and C. J. Batty: Nucl. Phys. A **579**, 518 (1994)
62. A. Cieplý, E. Friedman, A. Gal, and J. Mareš: Nucl. Phys. A **696**, 173 (2001)
63. N. Kaiser, P. B. Siegel, and W. Weise: Nucl. Phys. A **594**, 325 (1995)
64. T. Waas, N. Kaiser, and W. Weise: Phys. Lett. B **365**, 12 (1996); **379**, 34 (1996)
65. A. Ramos and E. Oset: Nucl. Phys. A **671**, 481 (2000)
66. A. Baca, C. García-Recio, and J. Nieves: Nucl. Phys. A **673**, 335 (2000)
67. C. J. Batty, E. Friedman, and A. Gal: Phys. Rep. **287**, 385 (1997)

The Hyperon-Nucleon Interaction: Conventional Versus Effective Field Theory Approach

J. Haidenbauer[1], U.-G. Meißner[1,2], A. Nogga[1] and H. Polinder[1]

[1] Institut für Kernphysik (Theorie), Forschungszentrum Jülich, D-52425 Jülich,
Germany
j.haidenbauer@fz-juelich.de; u.meissner@fz-juelich.de;
a.nogga@fz-juelich.de; h.polinder@fz-juelich.de
[2] Helmholtz-Institut für Strahlen- und Kernphysik (Theorie), Universität Bonn,
Nußallee 14–16, D-53115 Bonn, Germany
u.meissner@fz-juelich.de

Abstract. Hyperon-nucleon interactions are presented that are derived either in the conventional meson-exchange picture or within leading order chiral effective field theory. The chiral potential consists of one-pseudoscalar-meson exchanges and non-derivative four-baryon contact terms. With regard to meson-exchange YN models we focus on the new potential of the Jülich group, whose most salient feature is that the contributions in the scalar–isoscalar (σ) and vector–isovector (ϱ) exchange channels are constrained by a microscopic model of correlated $\pi\pi$ and $K\bar{K}$ exchange.

1 Introduction

For several decades the meson-exchange picture provided the only practicable and systematic approach to the description of hadronic reactions in the low- and medium-energy regime. Specifically, for the fundamental nucleon-nucleon (NN) interaction rather precise quantitative results could be achieved with meson-exchange models [1, 2]. Moreover, utilizing for example SU(3)$_\mathrm{f}$ (flavor) symmetry or G-parity arguments, within the meson-exchange framework, interaction models for the hyperon-nucleon (YN) [3, 4, 5, 6, 7, 8, 9, 10] or nucleon-antinucleon (N$\bar{\mathrm{N}}$) [11] systems could be constructed consistently. However, over the last 10 years or so a new powerful tool has emerged, namely chiral perturbation theory or, generally speaking, effective field theory (EFT). The main advantage of this scheme is that there is an underlying power counting that allows to improve calculations systematically by going to higher orders and, at the same time, provides theoretical uncertainties. In addition, it is possible to derive two- and corresponding three-body forces as well as external current operators in a consistent way. For reviews we refer to [12, 13, 14].

J. Haidenbauer et al.: *The Hyperon-Nucleon Interaction: Conventional Versus Effective Field Theory Approach*, Lect. Notes Phys. **724**, 113–140 (2007)
DOI 10.1007/978-3-540-72039-3_4 © Springer-Verlag Berlin Heidelberg 2007

Recently the NN interaction has been described to a high precision using chiral EFT [15] (see also [16]). In that work, the power counting is applied to the NN potential, as originally proposed by Weinberg [17, 18]. The NN potential consists of pion exchanges and a series of contact interactions with an increasing number of derivatives to parameterize the shorter ranged part of the NN force. A regularized Lippmann-Schwinger equation is solved to calculate observable quantities. Note that in contrast to the original Weinberg scheme, the effective potential is made explicitly energy-independent as it is important for applications in few-nucleon systems (for details, see [19]).

Contrary to the NN system, there are very few investigations of the YN interaction using EFT. Hyperon and nucleon mass shifts in nuclear matter, using chiral perturbation theory, have been studied in [20]. These authors used a chiral interaction containing four-baryon contact terms and pseudoscalar-meson exchanges. Recently, the hypertriton and Λd scattering were investigated in the framework of an EFT with contact interactions only [21]. Korpa et al [22] performed a next-to-leading order (NLO) EFT analysis of YN scattering and hyperon mass shifts in nuclear matter. Their tree-level amplitude contains four-baryon contact terms; pseudoscalar-meson exchanges were not considered explicitly, but $SU(3)_f$ breaking by meson masses was modeled by incorporating dimension-two terms coming from one-pion exchange. The full scattering amplitude was calculated using the Kaplan-Savage-Wise resummation scheme [23]. The YN scattering data were described successfully for laboratory momenta below $200\,\mathrm{MeV}$, using 12 free parameters. Some aspects of strong ΛN scattering in EFT and its relation to various formulations of lattice QCD are discussed in [24]. Finally, in this context we note that first lattice QCD results on the YN interaction have appeared [25].

In this review we describe a recent application of the scheme used in [15] to the YN interaction by the Bonn-Jülich group [26]. Analogous to the NN potential, at leading order (LO) in the power counting, the YN potential consists of pseudoscalar-meson (Goldstone boson) exchanges and of four-baryon contact terms, where each of these two contributions is constrained via $SU(3)_f$ symmetry. The results achieved by us within this approach are confronted with the available YN data and they are also compared with predictions of a new conventional meson-exchange YN model, developed likewise by the Jülich group [9], whose most salient feature is that the contributions in the scalar–isoscalar (σ) and vector–isovector (ϱ) exchange channels are constrained by a microscopic model of correlated $\pi\pi$ and $K\overline{K}$ exchange. Results of the Nijmegen YN model NSC97f [7] are presented too.

The contents of this review are as follows. In Sect. 2 we discuss some general properties of the coupled ΛN and ΣN systems. We also introduce the coupled-channels Lippmann-Schwinger equation that is solved for obtaining the reaction amplitude. The effective potential in leading order chiral EFT is developed in Sect. 3. Here we first give a brief recollection of the underlying power counting for the effective potential and then investigate the $SU(3)_f$ structure of the four-baryon contact interactions. The lowest order $SU(3)_f$-

invariant contributions from pseudoscalar meson exchange are derived too. Some general remarks about meson-exchange potentials of the YN interaction are given in Sect. 4. We also provide a more specific description of the new meson-exchange potential of the Jülich group [9], where we focus on the utilized model of correlated $\pi\pi$ and $K\overline{K}$ exchange. Results of both interactions for low-energy YN cross sections are presented in Sect. 5. We show the empirical and calculated total cross sections, differential cross sections and give the values for the scattering lengths. Also, predictions for some YN phase shifts are presented and results for binding energies of light hypernuclei are listed. The review closes with a summary and an outlook for future investigations.

2 The Scattering Equation

In the meson-meson and meson-baryon sector, chiral interactions can be treated perturbatively in powers of a low-energy scale (chiral perturbation theory). This is not the case for the baryon-baryon sector, otherwise there could be no bound states, such as the deuteron. Weinberg [18] realized that an additional scale arises from intermediate states with only two nucleons, which requires a modification of the power counting. He proposed to apply the techniques of chiral perturbation theory to derive an effective potential, V, and not directly the scattering amplitude. This effective potential is defined as the sum of all irreducible diagrams. The effective potential V is then put into a Lippmann-Schwinger equation to obtain the reaction or scattering amplitude,

$$T = V + VGT,\tag{1}$$

where G is the non-relativistic free two-body Green's function. Solving the scattering equation (1) also implies that the reaction amplitude T fulfills two-body unitarity.

Treating the Lippmann-Schwinger equation for the YN system is more involved than for the NN system. Since the mass difference between the Λ and Σ hyperons is only about $75\,\mathrm{MeV}$ the possible coupling between the ΛN and ΣN systems needs to be taken into account. Moreover, for a sensible comparison of the results with experiments it is preferable to solve the scattering equation in the particle basis because then the Coulomb interaction in the charged channels can be incorporated. Here we use the method originally introduced by Vincent and Phatak [27] that was e.g. also applied in the EFT studies of the NN interaction [28]. Furthermore, the particle basis allows to implement the correct physical thresholds of the various ΣN channels. To facilitate the latter aspect we also use relativistic kinematics for relating the total energy $\sqrt{s}$ to the c.m. momenta in the various YN channels in the actual calculations, cf. [9]. Note that the interaction potentials themselves are calculated in the isospin basis.

The concrete particle channels that couple for a specific charge Q are

$$Q = +2 : \Sigma^+ p$$
$$Q = +1 : \Lambda p, \Sigma^+ n, \Sigma^0 p$$
$$Q = \ \ 0 : \Lambda n, \Sigma^0 n, \Sigma^- p$$
$$Q = -1 : \Sigma^- n \tag{2}$$

Therefore, e.g., for $Q = 0$ the quantities in (1) are then 3×3 matrices,

$$V = \begin{pmatrix} V_{\Lambda n \to \Lambda n} & V_{\Lambda n \to \Sigma^0 n} & V_{\Lambda n \to \Sigma^- p} \\ V_{\Sigma^0 n \to \Lambda n} & V_{\Sigma^0 n \to \Sigma^0 n} & V_{\Sigma^0 n \to \Sigma^- p} \\ V_{\Sigma^- p \to \Lambda n} & V_{\Sigma^- p \to \Sigma^0 n} & V_{\Sigma^- p \to \Sigma^- p} \end{pmatrix} , \tag{3}$$

and analogously for T while the Green's function is a diagonal matrix,

$$G = \begin{pmatrix} G_{\Lambda n} & 0 & 0 \\ 0 & G_{\Sigma^0 n} & 0 \\ 0 & 0 & G_{\Sigma^- p} \end{pmatrix} . \tag{4}$$

Explicitly, G_i is given by

$$G_i = \left[\frac{p_i^2 - \mathbf{p}'^2}{2\mu_i} + i\varepsilon \right]^{-1} , \tag{5}$$

where $\mu_i = M_{Y_i} M_{N_i} / (M_{Y_i} + M_{N_i})$ is the reduced mass and $\mathbf{p}'$ the c.m. momentum in the intermediate $Y_i N_i$ channel. $p_i = p_i(\sqrt{s})$ denotes the modulus of the on-shell momentum in the intermediate $Y_i N_i$ state defined by $\sqrt{s} = \sqrt{M_{Y_i}^2 + p_i^2} + \sqrt{M_{N_i}^2 + p_i^2}$.

3 Hyperon-Nucleon Potential Based on Effective Field Theory

In this Section, we construct in some detail the effective chiral YN potential at leading order in the (modified) Weinberg power counting. This power counting is briefly recalled first. Then, we construct the minimal set of non-derivative four-baryon interactions and derive the formulae for the one-Goldstone-boson-exchange contributions.

3.1 Power Counting

In our work [26] we apply the power counting to the effective YN potential V which is then injected into a Lippmann-Schwinger equation (1) to generate the bound and scattering states. The various terms in the effective potential are ordered according to

$$V \equiv V(Q, g, \mu) = \sum_\nu Q^\nu \, \mathcal{V}_\nu(Q/\mu, g) , \qquad (6)$$

where Q is the soft scale (either a baryon three-momentum, a Goldstone boson four-momentum or a Goldstone boson mass), g is a generic symbol for the pertinent low–energy constants, μ a regularization scale, $\mathcal{V}_\nu$ is a function of order one, and $\nu \geq 0$ is the chiral power. It can be expressed as [14]

$$\nu = 2 - B + 2L + \sum_i v_i \, \Delta_i \ ,$$

$$\Delta_i = d_i + \frac{1}{2} b_i - 2 \ , \qquad (7)$$

with B the number of incoming (outgoing) baryon fields, L counts the number of Goldstone boson loops, and v_i is the number of vertices with dimension Δ_i. The vertex dimension is expressed in terms of derivatives (or Goldstone boson masses) d_i and the number of internal baryon fields b_i at the vertex under consideration. The LO potential is given by $\nu = 0$, with $B = 2$, $L = 0$ and $\Delta_i = 0$. Using (7) it is easy to see that this condition is fulfilled for two types of interactions – a) non-derivative four-baryon contact terms with $b_i = 4$ and $d_i = 0$ and b) one-meson exchange diagrams with the leading meson-baryon derivative vertices allowed by chiral symmetry ($b_i = 2, d_i = 1$). At LO, the effective potential is entirely given by these two types of contributions.

3.2 The Four-Baryon Contact Terms

Let us start with briefly recalling the situation for the NN interactions. The LO contact term for the NN interactions is given by e.g. [17, 19]

$$\mathcal{L} = C_i \left(\bar{N} \Gamma_i N \right) \left(\bar{N} \Gamma_i N \right) , \qquad (8)$$

where Γ_i are the usual elements of the Clifford algebra [29]

$$\Gamma_1 = 1 \ , \ \Gamma_2 = \gamma^\mu \ , \ \Gamma_3 = \sigma^{\mu\nu} \ , \ \Gamma_4 = \gamma^\mu \gamma_5 \ , \ \Gamma_5 = \gamma_5 \ , \qquad (9)$$

N are the Dirac spinors of the nucleons and C_i are the so-called low-energy constants (LECs). The small components of the nucleon spinors do not contribute to the LO contact interactions. Considering the large components only, the LO contact term, (8), becomes

$$\mathcal{L} = -\frac{1}{2} C_{\mathrm{S}} \left(\varphi_N^\dagger \varphi_N \right) \left(\varphi_N^\dagger \varphi_N \right) - \frac{1}{2} C_{\mathrm{T}} \left(\varphi_N^\dagger \boldsymbol{\sigma} \varphi_N \right) \left(\varphi_N^\dagger \boldsymbol{\sigma} \varphi_N \right) , \qquad (10)$$

where φ_N denotes the large component of the Dirac spinor and C_{S} and C_{T} are the LECs that need to be determined by fitting to the experimental data.

In the case of the YN interaction we will consider a similar but $SU(3)_{\mathrm{f}}$ invariant coupling. The LO contact terms for the octet baryon-baryon interactions, that are Hermitian and invariant under Lorentz transformations, are given by the $SU(3)_{\mathrm{f}}$ invariants,

$$\mathcal{L}^1 = C_i^1 \left\langle \bar{B}_a \bar{B}_b \left(\Gamma_i B\right)_b \left(\Gamma_i B\right)_a \right\rangle , \quad \mathcal{L}^2 = C_i^2 \left\langle \bar{B}_a \left(\Gamma_i B\right)_a \bar{B}_b \left(\Gamma_i B\right)_b \right\rangle ,$$

$$\mathcal{L}^3 = C_i^3 \left\langle \bar{B}_a \left(\Gamma_i B\right)_a \right\rangle \left\langle \bar{B}_b \left(\Gamma_i B\right)_b \right\rangle , \quad \mathcal{L}^4 = C_i^4 \left\langle \bar{B}_a \bar{B}_b \left(\Gamma_i B\right)_a \left(\Gamma_i B\right)_b \right\rangle ,$$

$$\mathcal{L}^5 = C_i^5 \left\langle \bar{B}_a \left(\Gamma_i B\right)_b \bar{B}_b \left(\Gamma_i B\right)_a \right\rangle , \quad \mathcal{L}^6 = C_i^6 \left\langle \bar{B}_a \left(\Gamma_i B\right)_b \right\rangle \left\langle \bar{B}_b \left(\Gamma_i B\right)_a \right\rangle ,$$

$$\mathcal{L}^7 = C_i^7 \left\langle \bar{B}_a \left(\Gamma_i B\right)_a \left(\Gamma_i B\right)_b \bar{B}_b \right\rangle , \quad \mathcal{L}^8 = C_i^8 \left\langle \bar{B}_a \left(\Gamma_i B\right)_b \left(\Gamma_i B\right)_a \bar{B}_b \right\rangle ,$$

$$\mathcal{L}^9 = C_i^9 \left\langle \bar{B}_a \bar{B}_b \right\rangle \left\langle \left(\Gamma_i B\right)_a \left(\Gamma_i B\right)_b \right\rangle . \tag{11}$$

Here a and b denote the Dirac indices of the particles, B is the usual irreducible octet representation of $SU(3)_f$ given by

$$B = \begin{pmatrix} \frac{\Sigma^0}{\sqrt{2}} + \frac{\Lambda}{\sqrt{6}} & \Sigma^+ & p \\ \Sigma^- & -\frac{\Sigma^0}{\sqrt{2}} + \frac{\Lambda}{\sqrt{6}} & n \\ -\Xi^- & \Xi^0 & -\frac{2\Lambda}{\sqrt{6}} \end{pmatrix} , \tag{12}$$

and the brackets in (11) denote taking the trace in the three-dimensional flavor space. The Clifford algebra elements are here actually diagonal 3×3 matrices in flavor space. Term 9 in (11) can be eliminated using a Cayley-Hamilton identity

$$-\left\langle \bar{B}_a \bar{B}_b \left(\Gamma_i B\right)_a \left(\Gamma_i B\right)_b \right\rangle + \left\langle \bar{B}_a \bar{B}_b \left(\Gamma_i B\right)_b \left(\Gamma_i B\right)_a \right\rangle$$

$$-\frac{1}{2} \left\langle \bar{B}_a \left(\Gamma_i B\right)_b \bar{B}_b \left(\Gamma_i B\right)_a \right\rangle + \frac{1}{2} \left\langle \bar{B}_a \left(\Gamma_i B\right)_a \bar{B}_b \left(\Gamma_i B\right)_b \right\rangle$$

$$= \frac{1}{2} \left\langle \bar{B}_a \left(\Gamma_i B\right)_a \right\rangle \left\langle \bar{B}_b \left(\Gamma_i B\right)_b \right\rangle - \frac{1}{2} \left\langle \bar{B}_a \left(\Gamma_i B\right)_b \right\rangle \left\langle \bar{B}_b \left(\Gamma_i B\right)_a \right\rangle$$

$$-\frac{1}{2} \left\langle \bar{B}_a \bar{B}_b \right\rangle \left\langle \left(\Gamma_i B\right)_a \left(\Gamma_i B\right)_b \right\rangle . \tag{13}$$

Making use of the trace property $\langle AB \rangle = \langle BA \rangle$, we see that the terms 7 and 8 in (11) are equivalent to the terms 1 and 4 respectively. Also making use of the Fierz theorem, see e.g. [30], one can show that the terms 4, 5 and 6 are equivalent to the terms 1, 2 and 3, respectively. So, the minimal set of non-derivative four baryon contact interactions is given by $\mathcal{L}^1$, $\mathcal{L}^2$ and $\mathcal{L}^3$. Writing these interaction Lagrangians explicitly in the isospin basis we find for the NN and YN interactions

$$\mathcal{L}^1 = C_i^1 \left\{ \frac{1}{6} \left[5 \left(\bar{\Lambda} \Gamma_i \Lambda\right) \left(\bar{N} \Gamma_i N\right) - 4 \left(\bar{N} \Gamma_i \Lambda\right) \left(\bar{\Lambda} \Gamma_i N\right) \right] \right.$$

$$+ \frac{1}{2} \left[\left(\bar{\boldsymbol{\Sigma}} \cdot \Gamma_i \boldsymbol{\Sigma}\right) \left(\bar{N} \Gamma_i N\right) + i \left(\bar{\boldsymbol{\Sigma}} \times \Gamma_i \boldsymbol{\Sigma}\right) \cdot \left(\bar{N} \boldsymbol{\tau} \Gamma_i N\right) \right]$$

$$+ \frac{1}{\sqrt{12}} \left[\left\{ \left(\bar{N} \boldsymbol{\tau} \Gamma_i N\right) \cdot \left(\bar{\Lambda} \Gamma_i \boldsymbol{\Sigma}\right) + \text{H.c.} \right\} \right.$$

$$\left. \left. - 2 \left\{ \left(\bar{N} \Gamma_i \boldsymbol{\Sigma}\right) \cdot \left(\bar{\Lambda} \boldsymbol{\tau} \Gamma_i N\right) + \text{H.c.} \right\} \right] \right\} ,$$

$$\mathcal{L}^2 = C_i^2 \left\{ \frac{1}{3} \left[4 \left(\bar{\Lambda} \Gamma_i \Lambda\right) \left(\bar{N} \Gamma_i N\right) + \left(\bar{N} \Gamma_i \Lambda\right) \left(\bar{\Lambda} \Gamma_i N\right) \right] \right.$$

$$+ \left[(\bar{N}\Gamma_i \boldsymbol{\Sigma}) \cdot (\bar{\boldsymbol{\Sigma}}\Gamma_i N) + i\, (\bar{N}\Gamma_i \boldsymbol{\Sigma}) \cdot (\bar{\boldsymbol{\Sigma}} \times \boldsymbol{\tau}\Gamma_i N) \right]$$

$$+ \frac{1}{\sqrt{3}} \left[(\bar{N}\Gamma_i \boldsymbol{\Sigma}) \cdot (\bar{\Lambda}\boldsymbol{\tau}\Gamma_i N) + \text{H.c.} \right] + (\bar{N}\Gamma_i N)(\bar{N}\Gamma_i N) \Big\} ,$$

$$\mathcal{L}^3 = C_i^3 \left\{ 2 \left(\bar{\Lambda}\Gamma_i \Lambda \right) (\bar{N}\Gamma_i N) + 2 \left(\bar{\boldsymbol{\Sigma}} \cdot \Gamma_i \boldsymbol{\Sigma} \right) (\bar{N}\Gamma_i N) + (\bar{N}\Gamma_i N)(\bar{N}\Gamma_i N) \right\} . \tag{14}$$

Here H.c. denotes the Hermitian conjugate of the specific term. Also Λ is an isoscalar, N and Ξ are isospinors and Σ is an isovector:

$$N = \begin{pmatrix} p \\ n \end{pmatrix}, \quad \Xi = \begin{pmatrix} \Xi^0 \\ \Xi^- \end{pmatrix}, \quad \boldsymbol{\Sigma} = \begin{pmatrix} \Sigma^+ \\ \Sigma^0 \\ \Sigma^- \end{pmatrix} . \tag{15}$$

The LO YN contact terms given by these Lagrangians are shown diagrammatically in Fig. 1. Considering again only the large components of the Dirac spinors, similar to (10), we arrive at six contact constants (C_S^1, C_T^1, C_S^2, C_T^2, C_S^3 and C_T^3) for the interactions in the various BB → BB channels. The LO contact potentials resulting from the above Lagrangians have the form

$$V^{BB \to BB} = C_S^{BB \to BB} + C_T^{BB \to BB}\, \boldsymbol{\sigma_1} \cdot \boldsymbol{\sigma_2} . \tag{16}$$

Projecting the LO contact potential on the partial waves, for details see, e.g. [31], one finds the following contributions. The NN partial wave potentials are

$$V_{1S0}^{NN} = 4\pi \left[2 \left(C_S^2 - 3C_T^2 \right) + 2 \left(C_S^3 - 3C_T^3 \right) \right] = V^{27} ,$$

$$V_{3S1}^{NN} = 4\pi \left[2 \left(C_S^2 + C_T^2 \right) + 2 \left(C_S^3 + C_T^3 \right) \right] = V^{10^*} . \tag{17}$$

The partial wave potentials for $\Lambda N \to \Lambda N$ are

$$V_{1S0}^{\Lambda\Lambda} = 4\pi \left[\frac{1}{6} \left(C_S^1 - 3C_T^1 \right) + \frac{5}{3} \left(C_S^2 - 3C_T^2 \right) + 2 \left(C_S^3 - 3C_T^3 \right) \right]$$

$$= \frac{1}{10} \left(9V^{27} + V^{8s} \right) ,$$

$$V_{3S1}^{\Lambda\Lambda} = 4\pi \left[\frac{3}{2} \left(C_S^1 + C_T^1 \right) + \left(C_S^2 + C_T^2 \right) + 2 \left(C_S^3 + C_T^3 \right) \right]$$

$$= \frac{1}{2} \left(V^{8a} + V^{10^*} \right) , \tag{18}$$

Fig. 1. Lowest order contact terms for hyperon-nucleon interactions

where here and in the following we introduced the shorthand notation "$\Lambda\Lambda$" instead of $\Lambda N \to \Lambda N$, etc., for labelling the interaction potentials and the corresponding contact terms. For isospin-3/2 $\Sigma N \to \Sigma N$ one gets

$$V_{1S0}^{\Sigma\Sigma} = 4\pi \left[2 \left(C_S^2 - 3C_T^2\right) + 2 \left(C_S^3 - 3C_T^3\right)\right] = V^{27} ,$$
$$V_{3S1}^{\Sigma\Sigma} = 4\pi \left[-2 \left(C_S^2 + C_T^2\right) + 2 \left(C_S^3 + C_T^3\right)\right] = V^{10} , \qquad (19)$$

for isospin-1/2 $\Sigma N \to \Sigma N$

$$\widetilde{V}_{1S0}^{\Sigma\Sigma} = 4\pi \left[\frac{3}{2} \left(C_S^1 - 3C_T^1\right) - \left(C_S^2 - 3C_T^2\right) + 2 \left(C_S^3 - 3C_T^3\right)\right]$$
$$= \frac{1}{10} \left(V^{27} + 9V^{8s}\right) ,$$
$$\widetilde{V}_{3S1}^{\Sigma\Sigma} = 4\pi \left[\frac{3}{2} \left(C_S^1 + C_T^1\right) + \left(C_S^2 + C_T^2\right) + 2 \left(C_S^3 + C_T^3\right)\right]$$
$$= \frac{1}{2} \left(V^{8a} + V^{10^*}\right) , \qquad (20)$$

and for $\Lambda N \to \Sigma N$

$$V_{1S0}^{\Lambda\Sigma} = 4\pi \left[\frac{1}{2} \left(C_S^1 - 3C_T^1\right) - \left(C_S^2 - 3C_T^2\right)\right] = \frac{3}{10} \left(-V^{27} + V^{8s}\right) ,$$
$$V_{3S1}^{\Lambda\Sigma} = 4\pi \left[-\frac{3}{2} \left(C_S^1 + C_T^1\right) + \left(C_S^2 + C_T^2\right)\right] = \frac{1}{2} \left(-V^{8a} + V^{10^*}\right) . \quad (21)$$

The last relations in the previous (17)–(21) give explicitly the SU(3)$_f$ representation of the potentials, see [32, 33]. We note that only 5 of the $\{8\} \times \{8\} = \{27\} + \{10\} + \{10^*\} + \{8\}_s + \{8\}_a + \{1\}$ irreducible representations are relevant for NN and YN interactions, since the $\{1\}$ occurs only in the isospin zero $\Lambda\Lambda$, ΞN and $\Sigma\Sigma$ channels. Equivalently, the six contact terms, C_S^1, C_T^1, C_S^2, C_T^2, C_S^3, C_T^3, enter the NN and YN potentials in only 5 different combinations. These 5 contact terms need to be determined by a fit to the experimental data. Since the NN data can not be described well with a LO EFT, see [17, 34], we will not consider the NN interaction explicitly. Therefore, we are left with the YN partial wave potentials

$$V_{1S0}^{\Lambda\Lambda} = C_{1S0}^{\Lambda\Lambda} , \qquad\qquad V_{3S1}^{\Lambda\Lambda} = C_{3S1}^{\Lambda\Lambda} ,$$
$$V_{1S0}^{\Sigma\Sigma} = C_{1S0}^{\Sigma\Sigma} , \qquad\qquad V_{3S1}^{\Sigma\Sigma} = C_{3S1}^{\Sigma\Sigma} ,$$
$$\widetilde{V}_{1S0}^{\Sigma\Sigma} = 9C_{1S0}^{\Lambda\Lambda} - 8C_{1S0}^{\Sigma\Sigma} , \quad \widetilde{V}_{3S1}^{\Sigma\Sigma} = C_{3S1}^{\Lambda\Lambda} , \qquad\qquad (22)$$
$$V_{1S0}^{\Lambda\Sigma} = 3 \left(C_{1S0}^{\Lambda\Lambda} - C_{1S0}^{\Sigma\Sigma}\right) , \; V_{3S1}^{\Lambda\Sigma} = C_{3S1}^{\Lambda\Sigma} .$$

We have chosen to search for $C_{1S0}^{\Lambda\Lambda}$, $C_{3S1}^{\Lambda\Lambda}$, $C_{1S0}^{\Sigma\Sigma}$, $C_{3S1}^{\Sigma\Sigma}$, and $C_{3S1}^{\Lambda\Sigma}$ in the fitting procedure. The other partial wave potentials are then fixed by SU(3)$_f$ symmetry.

3.3 One Pseudoscalar-Meson Exchange

The lowest order $SU(3)_f$-invariant pseudoscalar-meson–baryon interaction Lagrangian is given by (see, e.g. [35]),

$$\mathcal{L} = \left\langle i\,\bar{B}\gamma^\mu D_\mu B - M_0 \bar{B}B + \frac{D}{2}\bar{B}\gamma^\mu\gamma_5\{u_\mu, B\} + \frac{F}{2}\bar{B}\gamma^\mu\gamma_5[u_\mu, B]\right\rangle, \quad (23)$$

with M_0 the octet baryon mass in the chiral limit. There are two possibilities for coupling the axial vector u_μ to the baryon bilinear. The conventional coupling constants F and D, used here, satisfy the relation $F+D = g_A \simeq 1.26$. The axial-vector strength g_A is measured in neutron β–decay. The covariant derivative acting on the baryons is

$$D_\mu B = \partial_\mu B + [\Gamma_\mu, B]\,,$$

$$\Gamma_\mu = \frac{1}{2}\left[u^\dagger \partial_\mu u + u \partial_\mu u^\dagger\right]\,,$$

$$u^2 = U = \exp(2i\,P/\sqrt{2}F_\pi)\,,$$

$$u_\mu = i\,u^\dagger \partial_\mu U u^\dagger\,, \quad (24)$$

where F_π is the weak pion decay constant, $F_\pi = 92.4\,\text{MeV}$, and P is the irreducible octet representation of $SU(3)_f$ for the pseudoscalar mesons (the Goldstone bosons)

$$P = \begin{pmatrix} \frac{\pi^0}{\sqrt{2}} + \frac{\eta}{\sqrt{6}} & \pi^+ & K^+ \\ \pi^- & \frac{-\pi^0}{\sqrt{2}} + \frac{\eta}{\sqrt{6}} & K^0 \\ K^- & \bar{K}^0 & -\frac{2\eta}{\sqrt{6}} \end{pmatrix}. \quad (25)$$

Symmetry breaking in the decay constants, e.g. $F_\pi \neq F_K$, formally appears at NLO and will not be considered in the following. Writing the interaction Lagrangian explicitly in the isospin basis, we find

$$\begin{aligned} \mathcal{L} = &-f_{NN\pi}\,\bar{N}\gamma^\mu\gamma_5\boldsymbol{\tau}N \cdot \partial_\mu\boldsymbol{\pi} \\ &+i\,f_{\Sigma\Sigma\pi}\,\bar{\boldsymbol{\Sigma}}\gamma^\mu\gamma_5 \times \boldsymbol{\Sigma} \cdot \partial_\mu\boldsymbol{\pi} \\ &-f_{\Lambda\Sigma\pi}\left[\bar{\Lambda}\gamma^\mu\gamma_5\boldsymbol{\Sigma} + \bar{\boldsymbol{\Sigma}}\gamma^\mu\gamma_5\Lambda\right] \cdot \partial_\mu\boldsymbol{\pi} \\ &-f_{\Xi\Xi\pi}\,\bar{\Xi}\gamma^\mu\gamma_5\boldsymbol{\tau}\Xi \cdot \partial_\mu\boldsymbol{\pi} \\ &-f_{\Lambda NK}\left[\bar{N}\gamma^\mu\gamma_5\Lambda\partial_\mu K + \bar{\Lambda}\gamma^\mu\gamma_5 N\partial_\mu K^\dagger\right] \\ &-f_{\Xi\Lambda K}\left[\bar{\Xi}\gamma^\mu\gamma_5\Lambda\partial_\mu K_c + \bar{\Lambda}\gamma^\mu\gamma_5\Xi\partial_\mu K_c^\dagger\right] \\ &-f_{\Sigma NK}\left[\bar{\boldsymbol{\Sigma}} \cdot \gamma^\mu\gamma_5\partial_\mu K^\dagger\boldsymbol{\tau}N + \bar{N}\gamma^\mu\gamma_5\boldsymbol{\tau}\partial_\mu K \cdot \boldsymbol{\Sigma}\right] \\ &-f_{\Sigma\Xi K}\left[\bar{\boldsymbol{\Sigma}} \cdot \gamma^\mu\gamma_5\partial_\mu K_c^\dagger\boldsymbol{\tau}\Xi + \bar{\Xi}\gamma^\mu\gamma_5\boldsymbol{\tau}\partial_\mu K_c \cdot \boldsymbol{\Sigma}\right] \\ &-f_{NN\eta_8}\bar{N}\gamma^\mu\gamma_5 N\partial_\mu\eta \\ &-f_{\Lambda\Lambda\eta_8}\bar{\Lambda}\gamma^\mu\gamma_5\Lambda\partial_\mu\eta \\ &-f_{\Sigma\Sigma\eta_8}\bar{\boldsymbol{\Sigma}} \cdot \gamma^\mu\gamma_5\boldsymbol{\Sigma}\partial_\mu\eta \\ &-f_{\Xi\Xi\eta_8}\bar{\Xi}\gamma^\mu\gamma_5\Xi\partial_\mu\eta\,. \end{aligned} \quad (26)$$

Here η is an isoscalar, K and K_c are isospin doublets

$$K = \begin{pmatrix} K^+ \\ K^0 \end{pmatrix} , \quad K_c = \begin{pmatrix} \bar{K}^0 \\ -K^- \end{pmatrix} , \tag{27}$$

and π is an isovector. The phases of the isovectors Σ and π are chosen such that [32]

$$\Sigma \cdot \pi = \Sigma^+ \pi^- + \Sigma^0 \pi^0 + \Sigma^- \pi^+ . \tag{28}$$

The interaction Lagrangian in (26) is invariant under $SU(3)_f$ transformations if the various coupling constants are expressed in terms of the coupling constant $f \equiv g_A/2F_\pi$ and the $F/(F+D)$-ratio α as [32],

$$
\begin{array}{llll}
f_{NN\pi} = f , & f_{NN\eta_8} = \frac{1}{\sqrt{3}}(4\alpha - 1)f , & f_{\Lambda NK} = -\frac{1}{\sqrt{3}}(1 + 2\alpha)f , \\
f_{\Xi\Xi\pi} = -(1 - 2\alpha)f , & f_{\Xi\Xi\eta_8} = -\frac{1}{\sqrt{3}}(1 + 2\alpha)f , & f_{\Xi\Lambda K} = \frac{1}{\sqrt{3}}(4\alpha - 1)f , \\
f_{\Lambda\Sigma\pi} = \frac{2}{\sqrt{3}}(1 - \alpha)f , & f_{\Sigma\Sigma\eta_8} = \frac{2}{\sqrt{3}}(1 - \alpha)f , & f_{\Sigma NK} = (1 - 2\alpha)f , \\
f_{\Sigma\Sigma\pi} = 2\alpha f , & f_{\Lambda\Lambda\eta_8} = -\frac{2}{\sqrt{3}}(1 - \alpha)f , & f_{\Xi\Sigma K} = -f .
\end{array}
\tag{29}
$$

The spin-space part of the one-pseudoscalar-meson-exchange potential resulting from the interaction Lagrangian (26) is in leading order, similar to the static one-pion-exchange potential (recoil and relativistic corrections give higher order contributions) in [19],

$$V^{B_1 B_2 \to B_1' B_2'} = -f_{B_1 B_1' P} f_{B_2 B_2' P} \frac{(\sigma_1 \cdot k)(\sigma_2 \cdot k)}{k^2 + m_P^2} , \tag{30}$$

where $f_{B_1 B_1' P}$, $f_{B_2 B_2' P}$ are the appropriate coupling constants as given in (29) and m_P is the actual mass of the exchanged pseudoscalar meson. Thus, the explicit $SU(3)$ breaking reflected in the mass splitting between the pseudoscalar mesons is taken into account. With regard to the η meson we identified its coupling with the octet value, i.e. the one for η_8, in our investigation [26]. (We will come back to that issue below.) We defined the transferred and average momentum, k and q, in terms of the final and initial center-of-mass (c.m.) momenta of the baryons, p' and p, as

$$k = p' - p , \quad q = \frac{p' + p}{2} . \tag{31}$$

To find the complete LO one-pseudoscalar-meson-exchange potential one needs to multiply the potential in (30) with the isospin factors given in Table 1. Figure 2 shows the one-pseudoscalar-meson-exchange diagrams. Note that there is no contribution from one-pion exchange to the $\Lambda N \to \Lambda N$ potential due to isospin conservation. Indeed, the longest ranged contribution to this interaction is provided by (iterated) two-pion exchange via the process $\Lambda N \to \Sigma N \to \Lambda N$, generated by solving the Lippmann-Schwinger equation (1).

Table 1. The isospin factors for the various one-pseudoscalar-meson exchanges

Channel	Isospin	π	K	η
NN $\rightarrow$ NN	0	-3	0	1
	1	1	0	1
ΛN $\rightarrow$ ΛN	$\frac{1}{2}$	0	1	1
ΛN $\rightarrow$ ΣN	$\frac{1}{2}$	$-\sqrt{3}$	$-\sqrt{3}$	0
ΣN $\rightarrow$ ΣN	$\frac{1}{2}$	-2	-1	1
	$\frac{3}{2}$	1	2	1

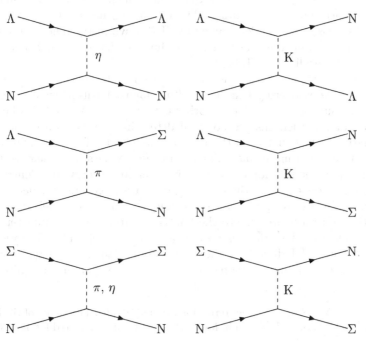

Fig. 2. One-pseudoscalar-meson-exchange diagrams for hyperon-nucleon interactions

3.4 Determination of the Low-Energy Constants

The chiral EFT potential in the Lippmann-Schwinger equation is cut off with the regulator function

$$f^{\Lambda}(p',p) = e^{-\left(p'^4 + p^4\right)/\Lambda^4} , \qquad (32)$$

in order to remove high-energy components of the baryon and pseudoscalar meson fields. For the cut-off Λ we consider values between 550 and 700 MeV.

This range is similar to the range used for chiral EFT NN interactions [15, 36, 37]. The range is limited from below by the mass of the pseudoscalar mesons; since we do a LO calculation we do not expect a large plateau (i.e. a practically stable χ^2 for varying Λ).

For the fitting procedure we considered the empirical low-energy total cross sections published in [38, 39, 40, 41] and the inelastic capture ratio at rest [42], in total 35 YN data [26]. These data are also commonly used for determining the parameters of meson-exchange models. The higher energy total cross sections and differential cross sections are then predictions of the LO chiral EFT, which contains five free parameters. The fits were done for fixed values of the cut–off mass (Λ) and of α, the pseudoscalar $F/(F+D)$ ratio. For the latter we used the SU(6) value: $\alpha = 0.4$. The five LECs $C_{1S0}^{\Lambda\Lambda}$, $C_{3S1}^{\Lambda\Lambda}$, $C_{1S0}^{\Sigma\Sigma}$, $C_{3S1}^{\Sigma\Sigma}$, and $C_{3S1}^{\Lambda\Sigma}$ in (22) were varied during the parameter search in order to fix the corresponding potentials. The interaction in the other YN partial waves (channels) are then determined by SU(3)$_f$ symmetry. The values of the contact terms obtained in the fitting procedure for cut–off values between 550 and 700 MeV, are listed in Table 2.

The fits were first done for the cut-off mass $\Lambda = 600$ MeV. We remark that the ΛN S-wave scattering lengths resulting for that cut-off were then kept fixed in the subsequent fits for the other cut–off values. We did this because the ΛN scattering lengths are not well determined by the scattering data. As a matter of fact, not even the relative magnitude of the ΛN triplet and singlet interaction can be constrained from the YN data, but their strengths play an important role for the hypertriton binding energy [6]. Contrary to the NN case, see, e.g. [34], the contact terms are in general not determined by a specific phase shift, because of the coupled particle channels in the YN interaction. Furthermore, due to the limited accuracy and incompleteness of the YN scattering data there is no partial wave analysis. Therefore we have fitted the chiral EFT directly to the cross sections. It is reassuring to see that the contact terms found in the parameter search are of similar magnitude as

Table 2. The YN S-wave contact terms for various cut–offs. The values of the LECs are in 10^4 GeV^{-2}; the values of Λ in MeV. χ^2 is the total chi squared for 35 YN data

Λ	550	600	650	700
$C_{1S0}^{\Lambda\Lambda}$	$-.0466$	$-.0403$	$-.0322$	$-.0304$
$C_{3S1}^{\Lambda\Lambda}$	$-.0222$	$-.0163$	$-.0097$	$-.0022$
$C_{1S0}^{\Sigma\Sigma}$	$-.0766$	$-.0763$	$-.0757$	$-.0744$
$C_{3S1}^{\Sigma\Sigma}$	$.2336$	$.2391$	$.2392$	$.2501$
$C_{3S1}^{\Lambda\Sigma}$	$-.0016$	$-.0019$	$.0000$	$.0035$
χ^2	29.6	28.3	30.3	34.6

those obtained in the application of chiral EFT to the NN interaction and, specifically, they are of natural size [14].

Note that we actually studied the dependence of our results on the pseudoscalar $F/(F + D)$ ratio α by varying it within a range of 10 percent; after refitting the contact terms we basically found an equally good description of the empirical data. An uncertainty in our calculation is the value of the η coupling, since we identified the physical η with the octet η as mentioned above. Therefore, we varied the η coupling between zero and its octet value, but we found very little influence on the description of the data (in fact, inclusion of the η leads to a better plateau of the χ^2 in the cut-off range considered).

4 Hyperon-Nucleon Models Based on the Conventional Meson-Exchange Picture

In the construction of conventional meson-exchange models of the YN interaction usually one likewise assumes $SU(3)_f$ symmetry for the occurring coupling constants, and in some cases even the $SU(6)$ symmetry of the quark model [4, 5]. Indeed, in the derivation of the meson-exchange contributions one follows essentially the same procedure as outlined in Sect. 3.3 for the case of pseudoscalar mesons and, therefore, we do not present it here explicitly. Details can be found in [4, 31, 43], for example. Of course, since besides the lowest pseudoscalar-meson multiplet also the exchanges of vector mesons (ϱ, ω, K^*), of scalar mesons (σ, ...), or even of axial-vector mesons ($a_1(1270)$, ...) [10] are included, one should keep in mind that the spin-space structure of the corresponding Lagrangians that enter into (23) differ and, accordingly, the final expressions for the corresponding contributions to the YN interaction potentials differ too. Also we want to emphasize that even for pseudoscalar mesons the final result for the interaction potentials differs, in general, from the expression given in (30). Contrary to the chiral EFT approach, recoil and relativistic corrections are often kept in meson-exchange models because no power counting rules are applied.

The major conceptual difference between the various meson-exchange models consists in the treatment of the scalar-meson sector. This simply reflects the fact that, unlike for pseudoscalar and vector mesons, so far there is no general agreement about who are the actual members of the lowest lying scalar meson $SU(3)$ multiplet. (For a thorough discussion on that issue and an overview of the extensive literature we refer the reader to [44, 45] and references therein.) Therefore, besides the question of the masses of the exchange particles it also remains unclear whether and how the relations for the coupling constants given in (29) should be applied. As a consequence, different prescriptions for describing the scalar sector, whose contributions play a crucial role in any baryon-baryon interaction at intermediate ranges, were adopted by the various authors who published meson-exchange models of the YN interaction.

For example, the Nijmegen group [3, 7, 10] views this interaction as being generated by genuine scalar-meson exchange. In their models NSC [3], NSC97 [7], and ESC04 [10] a scalar SU(3) nonet is exchanged — namely, two isospin-0 mesons (besides the $\varepsilon(760)$, the $S^*(975)$ ($f_0(980)$) in model NSC (NSC97, ESC04)), an isospin-1 meson (δ or $a_0(980)$) and an isospin-1/2 strange meson κ with a mass of 1000 MeV. A genuine scalar SU(3) nonet is also present in the so-called Ehime potential [8], where besides the $S^*(975)$ and δ (or $a_0(980)$) the $f_0(1581)$ and the $K_0^*(1429)$ are included. In addition the model incorporates two effective scalar-meson exchanges, $\sigma(484)$ and $\kappa(839)$, that stand for $(\pi\pi)_{I=0}$ and $(K\pi)_{I=1/2}$ correlations but are treated phenomenologically. In the older YN models of the Jülich group [4, 5] a σ (with a mass of ≈ 550 MeV) is included which is viewed as arising from correlated $\pi\pi$ exchange. In practice, however, the coupling strength of this fictitious σ to the baryons is treated as a free parameter and fitted to the data - a rather unsatisfactory feature of those models.

In the new meson-exchange YN potential presented recently by the Jülich group a different strategy is followed. Here, indeed, a microscopic model of correlated $\pi\pi$ and $K\overline{K}$ exchange is utilized to fix the contributions in the scalar-isoscalar (σ) and vector-isovector (ϱ) channels. The basic steps in evaluating these contributions are outlined in the next subsection. Besides correlated $\pi\pi$ and $K\overline{K}$ exchange the new YN model incorporates also the standard one-boson exchange contributions of the lowest pseudoscalar and vector meson multiplets with coupling constants determined by SU(3) symmetry relations (29). The so-called $F/(F+D)$ ratios, cf. Sect. 3.3, are fixed to $\alpha = 0.4$ ($\alpha = 1$) for the pseudoscalar (vector) meson multiplets by invoking SU(6) symmetry.

Let us mention for completeness that in meson-exchange models usually the Lippmann-Schwinger equation is not regularized by introducing a regulator function of the form (32) as in the EFT approach. For example, in case of the YN models of the Jülich group [4, 5, 9] convergence of the Lippmann-Schwinger equation is achieved by supplementing the interaction with form factors for each meson-baryon-baryon (xBB') vertex, cf. Sect. 2.3.3 of [4] for details. Those form factors are meant to take into account the extended hadron structure and are parametrized in the conventional monopole or dipole form, for example $F_{xBB'}(\mathbf{k}^2) = (\Lambda_{xBB'}^2 - m_x^2)/(\Lambda_{xBB'}^2 + \mathbf{k}^2)$, where $\mathbf{k}$ is the momentum transfer defined in (31), m_x is the mass of the exchanged meson and $\Lambda_{xBB'}$ is the so-called cut–off mass.

4.1 Model for Correlated $\pi\pi + K\overline{K}$ Exchange

The explicit derivation of the baryon-baryon interaction based on correlated $\pi\pi + K\overline{K}$ exchange is quite involved and we refer the interested reader to the work of Reuber et al [46] for the full details. Here we only describe briefly the principal steps of the derivation of the correlated $\pi\pi + K\overline{K}$ exchange potentials for the baryon-baryon amplitudes in the scalar-isoscalar (σ) and vector-isovector (ϱ) channels.

Based on a $\pi\pi-K\overline{K}$ amplitude the evaluation of the correlated $\pi\pi$ exchange process for the baryon-baryon reaction $A + B \rightarrow C + D$, cf. the cartoon in Fig. 3, can be done in two steps. Firstly the $A\overline{C} \rightarrow \pi\pi, K\overline{K}$ amplitude is determined in the pseudophysical region and then dispersion theory and unitarity are applied to connect this amplitude with the corresponding physical amplitudes in the $A + B \rightarrow C + D$ channel. In our concrete case A, B, etc. can be any combination of the baryons N, Λ, Σ, or Ξ.

The Born terms for the transitions $A\overline{C} \rightarrow \pi\pi, K\overline{K}$ include contributions from baryon exchange as well as ϱ-pole diagrams (cf. [47]). The correlations between the two pseudoscalar mesons are taken into account by means of a coupled channel ($\pi\pi$, $K\overline{K}$) model [47, 48] generated from s- and t-channel meson exchange Born terms. This model describes the empirical $\pi\pi$ phase shifts over a large energy range from threshold up to 1.3 GeV. The amplitudes for the $A\overline{C} \rightarrow \pi\pi$, $K\overline{K}$ transitions in the pseudophysical region are then obtained by solving a covariant scattering equation with full inclusion of the $\pi\pi$ - $K\overline{K}$ correlations. The parameters of the $A\overline{C} \rightarrow \pi\pi$, $K\overline{K}$ model, which are interrelated through SU(3) symmetry, are determined by fitting to the quasiempirical $N\overline{N} \rightarrow \pi\pi$ amplitudes in the pseudophysical region, $t \leq 4m_\pi^2$ [46], obtained by analytic continuation of the empirical πN and $\pi\pi$ data.

Assuming analyticity for the amplitudes dispersion relations can be formulated for the baryon-baryon amplitudes, which connect physical amplitudes in the s-channel with singularities and discontinuities of these amplitudes in the pseudophysical region of the t-channel processes for the $J^P = 0^+$ (σ) and 1^- (ϱ) channel:

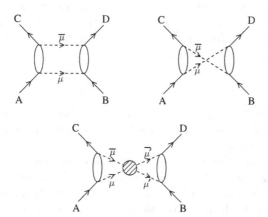

Fig. 3. Two-pion and two-kaon exchange in the baryon-baryon process $A + B \rightarrow C + D$. The *unshaded ellipse* denotes the direct coupling of the two pseudoscalar mesons $\mu\bar{\mu} = \pi\pi$, $K\overline{K}$, $\overline{K}K$ to the baryons without any correlation effects. The *shaded circle* in the lower diagram for the correlated exchange stands for the full off-shell amplitude of the process $\mu\bar{\mu} \rightarrow \mu'\bar{\mu}'$

$$V^{(0^+,1^-)}_{A,B \to C,D}(t) \propto \int_{4m_\pi^2}^\infty dt' \frac{\text{Im}V^{(0^+,1^-)}_{A,\overline{C} \to \overline{B},D}(t')}{t'-t}, \quad t < 0. \tag{33}$$

Via unitarity relations the singularity structure of the baryon-baryon amplitudes for $\pi\pi$ and $K\overline{K}$ exchange are fixed by and can be written as products of the $A\overline{C} \to \pi\pi$, $K\overline{K}$ amplitudes

$$\text{Im}V^{(0^+,1^-)}_{A,\overline{C} \to \overline{B},D}(t') \propto \sum_{\alpha=\pi\pi,K\overline{K}} T^{*,(0^+,1^-)}_{A,\overline{C} \to \alpha} T^{(0^+,1^-)}_{\overline{B},D \to \alpha}. \tag{34}$$

Thus, from the $A\overline{C} \to \pi\pi$, $K\overline{K}$ amplitudes the spectral functions can be calculated

$$\rho^{(0^+,1^-)}_{A,B \to C,D}(t') \propto \sum_{\alpha=\pi\pi,K\overline{K}} T^{*,(0^+,1^-)}_{A,\overline{C} \to \alpha} T^{(0^+,1^-)}_{\overline{B},D \to \alpha} \tag{35}$$

which are then inserted into dispersion integrals to obtain the (on-shell) baryon-baryon interaction:

$$V^{(0^+,1^-)}_{A,B \to C,D}(t) \propto \int_{4m_\pi^2}^\infty dt' \frac{\rho^{(0^+,1^-)}_{A,B \to C,D}(t')}{t'-t}, \quad t < 0. \tag{36}$$

The spectral function (35) for the (0^+) σ-channel has only one component but the one for the (1^-) ϱ-channel consists of four linearly independent components, which reflects the more complicated spin structure of this channel [46]. Note that the amplitudes in (33) still contain the uncorrelated (upper diagrams in Fig. 3), as well as the correlated pieces (lower diagram). Thus, in order to obtain the contribution of the truly correlated $\pi\pi$ and $K\overline{K}$ exchange one must eliminate the former from the spectral function. This is done by calculating the spectral function generated by the Born term and subtracting it from the total spectral function:

$$\rho^{(0^+,1^-)} \longrightarrow \rho^{(0^+,1^-)} - \rho^{(0^+,1^-)}_{\text{Born}}. \tag{37}$$

We should mention that the uncorrelated contributions are included too. But they are generated automatically by solving the scattering equation (1) for the interaction potential.

Finally, let us mention that the spectral functions characterize both the strength and range of the interaction. Clearly, for sharp mass exchanges the spectral function becomes a δ-function at the appropriate mass.

For convenience in the concrete calculations the potential due to correlated $\pi\pi/K\overline{K}$ exchange is parametrized in terms of effective coupling strengths of (sharp mass) σ and ϱ exchanges. The interaction resulting from the exchange of a σ meson with mass m_σ between two $J^P = 1/2^+$ baryons A and B has the structure:

$$V^\sigma_{A,B \to A,B}(t) = g_{AA\sigma} g_{BB\sigma} \frac{F_\sigma^2(t)}{t - m_\sigma^2}, \tag{38}$$

where a form factor $F_\sigma(t)$ is applied at each vertex, taking into account the fact that the exchanged σ meson is not on its mass shell. The correlated potential as given in (33) can now be parameterized in terms of t-dependent strength functions $G_{AB \to AB}(t)$, so that for the σ case:

$$V^{(0^+)}_{A,B \to A,B}(t) = G^\sigma_{AB \to AB}(t) F_\sigma^2(t) \frac{1}{t - m_\sigma^2} \, . \tag{39}$$

The effective coupling constants are then defined as

$$g_{AA\sigma} g_{BB\sigma} \longrightarrow G^\sigma_{AB \to AB}(t) = \frac{(t - m_\sigma^2)}{\pi F_\sigma^2(t)} \int_{4m_\pi^2}^\infty \frac{\rho^{(0^+)}_{AB \to AB}(t')}{t' - t} \, \mathrm{d}t' \, . \tag{40}$$

In the concrete application one varies m_σ^2 in order to achieve that $G^\sigma_{AB \to AB}(t) \approx G^\sigma_{AB \to AB}$, i.e. that $G^\sigma_{AB \to AB}$ is indeed practically a constant. The form factor is parameterized by

$$F_\sigma(t) = \frac{\Lambda_\sigma^2}{\Lambda_\sigma^2 - t} \, , \tag{41}$$

with a cut–off mass Λ_σ assumed to be the same for both vertices. This form guarantees that the on-shell behaviour of the potential (which is fully determined by the dispersion integral) is not modified strongly as long as the energy is not too high.

Similar relations can be also derived for the correlated exchange in the isovector-vector channel [46], which in this case will involve vector as well as tensor coupling pieces.

4.2 Other Ingredients of the Jülich Meson-Exchange Hyperon-Nucleon Model

Besides the correlated $\pi\pi$ and $K\bar{K}$ exchange the new YN model of the Jülich group takes into account exchange diagrams involving the well-established lowest lying pseudoscalar and vector meson SU(3) octets. Following the philosophy of the original Jülich YN potential [4] the coupling constants in the pseudoscalar sector are fixed by strict SU(6) symmetry. In any case, this is also required for being consistent with the model of correlated $\pi\pi$ and $K\bar{K}$ exchange. The cut–off masses of the form factors (cf. discussion at the beginning of Sect. 4) belonging to the NN vertices are taken over from the full Bonn NN potential. The cut–off masses at the strange vertices are considered as open parameters though, in practice, their values are kept as close as possible to those found for the NN vertices.

In addition there are some other new ingredients in the present YN model as compared to the earlier Jülich models [4, 5]. First of all, we now take into account contributions from (scalar-isovector) $a_0(980)$ exchange. The a_0 meson is present in the original Bonn NN potential [1], and for consistency should also

be included in the YN model. Secondly, we consider the exchange of a strange
scalar meson, the κ, with mass $\sim 1000\,\mathrm{MeV}$. Let us emphasize, however, that
like in case of the σ meson these particles are not viewed as being members of
a scalar meson SU(3) multiplet, but rather as representations of strong meson-
meson correlations in the scalar–isovector ($\pi\eta$–$\mathrm{K\overline{K}}$) [47] and scalar–isospin-1/2
(πK) channels [48], respectively. In principle, their contributions can also be
evaluated along the lines of [46], however, for simplicity in the present model
they are effectively parameterized by one-boson-exchange diagrams with the
appropriate quantum numbers assuming the coupling constants to be free
parameters. Furthermore, the new model contains the exchange of an ω' with
a mass of $m_{\omega'} = 1120\,\mathrm{MeV}$ considered to be an effective parametrization
of short-range contributions from correlated $\pi\varrho$ exchange [49] in the vector-
isoscalar sector. Its inclusion allows to keep the coupling constants of the
genuine $\omega(782)$ meson to the baryons in line with their SU(3) values, cf. the
discussion in [9]. In the spirit of the EFT approach, we have also considered
a version of the YN potential in [9] where the κ exchange was substituted by
a local contact interaction.

Thus we have the following scenario: The long- and intermediate-range
part of the new meson-exchange YN interaction model is completely deter-

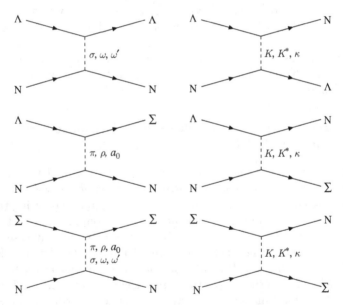

Fig. 4. Contributions to the meson-exchange YN model [9] in the ΛN and ΣN
channels and in the ΛN $\to$ ΣN transition. Note that only π, K, ω, and K* exchange
are considered as being due to genuine SU(3) mesons. The other contributions are
either fixed from correlated $\pi\pi$ and $\mathrm{K\overline{K}}$ exchange (σ, ϱ) or are viewed as an effective
parametrization of meson-meson correlations (a_0, κ, ω') in the corresponding spin-
isospin channels

mined by SU(6) constraints (for the pseudoscalar and, in general, also for the vector mesons) and by correlated $\pi\pi$ and $K\overline{K}$ exchange. The short-range part is viewed as being due to correlated meson-meson exchanges but in practice is parametrized phenomenologically in terms of one-boson-exchange contributions in specific spin-isospin channels. In particular, no SU(3) relations are imposed on the short-range part. This assumption is based on our observation that the contributions in the ϱ exchange channel as they result from correlated $\pi\pi$ and $K\overline{K}$ no longer fulfill strict SU(3) relations [46], but it also acknowledges the fact that at present there is no general agreement about who are the actual members of the lowest-lying scalar meson SU(3) multiplet, as already mentioned above. A graphical representation of all meson-exchange contributions that are included in the new YN model is given in Fig. 4.

5 Results and Discussion

In Fig. 5 we confront the results obtained from our YN interactions with the Λp, $\Sigma^+ p$, $\Sigma^- p$, $\Sigma^- p \to \Sigma^0 n$, and $\Sigma^- p \to \Lambda n$ data used in the fitting procedure. Here the solid curves correspond to the Jülich '04 meson-exchange model and the shaded band represents the results of the chiral EFT for the considered cut–off region. For reasons of comparison we also include the results of one of the meson-exchange models (NSC97f) of the Nijmegen group (dashed line) [7]. A detailed comparison between the experimental scattering data considered and the values found in the fitting procedure for the EFT interaction (for $\Lambda = 600\,\text{MeV}$) is given in Table 3. The differential cross sections are calculated in the usual way using the partial wave amplitudes, for details we refer to [4, 50]. The total cross sections are found by simply integrating the differential cross sections, except for the $\Sigma^+ p \to \Sigma^+ p$ and $\Sigma^- p \to \Sigma^- p$ channels. For those channels the experimental total cross sections were obtained via [40]

$$\sigma = \frac{2}{\cos\theta_{\max} - \cos\theta_{\min}} \int_{\cos\theta_{\min}}^{\cos\theta_{\max}} \frac{d\sigma(\theta)}{d\cos\theta} \, d\cos\theta , \qquad (42)$$

for various values of $\cos\theta_{\min}$ and $\cos\theta_{\max}$. Following [7], we use $\cos\theta_{\min} = -0.5$ and $\cos\theta_{\max} = 0.5$ in our calculations for the $\Sigma^+ p \to \Sigma^+ p$ and $\Sigma^- p \to \Sigma^- p$ cross sections, in order to stay as close as possible to the experimental procedure.

A good description of the low-energy YN scattering data has been obtained with the discussed meson-exchange models but also within the EFT approach in the considered cut–off region, as is documented in Tables 2 and 3 and Figs. 5 and 6.

Note that in the low-energy regime the cross sections are mainly given by the S-wave contribution, except for for the $\Lambda N \to \Sigma N$ cross section where

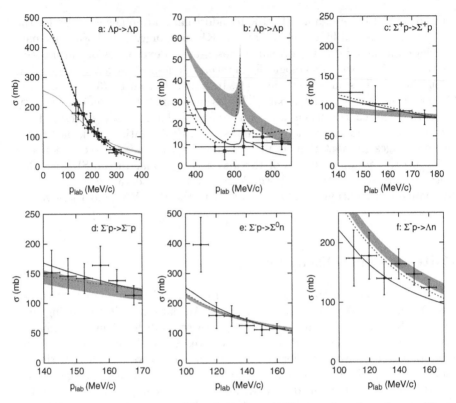

Fig. 5. "Total" cross section σ (as defined in (42)) as a function of p_{lab}. The experimental cross sections in (**a**) are taken from [38] (*open squares*) and [39] (*filled circles*), in (**b**) from [51] (*filled circles*) and [52] (*open squares*), in (**c**),(**d**) from [40] and in (**e**),(**f**) from [41]. The shaded band is the chiral EFT result for $\Lambda = 550, ..., 700$ MeV [26], the solid curve is the Jülich '04 model [9], and the dashed curve is the Nijmegen NSC97f potential [7]

the $^3D_1(\Lambda N) \leftrightarrow {}^3S_1(\Sigma N)$ transition provides the main contribution. Still all partial waves with total angular momentum $J \leq 2$ were included in the computation of the observables. The Λp cross sections show a clear cusp at the $\Sigma^+ n$ threshold, see Fig. 5b. This cusp is very pronounced for the EFT interaction, peaking at 60 mb, but also in case of the Nijmegen NSC97f model. It is hard to see this effect in the experimental data, since it occurs over a very narrow energy range. In case of the EFT interaction the predicted Λp cross section at higher energies is too large (cf. Fig. 5b), which is related to the problem that some LO phase shifts are too large at higher energies. Note that this is also the case for the NN interaction [34]. In a NLO calculation this problem will probably vanish. The differential cross sections at low energies, which have not been taken into account in the fitting procedure, are predicted well, see Fig. 7. The results of the meson-exchange models and of the chiral

Table 3. Comparison between the 35 experimental YN data and the results for the EFT interaction for the cut–off $\Lambda = 600\,\text{MeV}$. Momenta are in units of MeV and cross sections in mb. The achieved χ^2 is given for each reaction channel separately

$\Lambda p \to \Lambda p$ $\chi^2 = 7.5$			$\Lambda p \to \Lambda p$ $\chi^2 = 4.9$			$\Sigma^- p \to \Lambda n$ $\chi^2 = 5.5$		
p_{lab}^{Λ}	σ_{exp}[38]	σ_{the}	p_{lab}^{Λ}	σ_{exp}[39]	σ_{the}	$p_{\text{lab}}^{\Sigma^-}$	σ_{exp}[41]	σ_{the}
135	209±58	170.0	145	180±22	161.6	110	174±47	244.2
165	177±38	145.4	185	130±17	130.4	120	178±39	210.0
195	153±27	123.5	210	118±16	113.7	130	140±28	183.0
225	111±18	104.7	230	101±12	101.9	140	164±25	161.4
255	87 ±13	89.1	250	83 ±13	91.5	150	147±19	143.9
300	46 ±11	70.6	290	57 ±9	74.3	160	124±14	129.5

$\Sigma^+ p \to \Sigma^+ p$ $\chi^2 = 0.6$			$\Sigma^- p \to \Sigma^- p$ $\chi^2 = 2.4$			$\Sigma^- p \to \Sigma^0 n$ $\chi^2 = 7.0$		
$p_{\text{lab}}^{\Sigma^+}$	σ_{exp}[40]	σ_{the}	$p_{\text{lab}}^{\Sigma^-}$	σ_{exp}[40]	σ_{the}	$p_{\text{lab}}^{\Sigma^-}$	σ_{exp}[41]	σ_{the}
145	123±62	96.7	142.5	152±38	143.4	110	396±91	200.0
155	104±30	93.0	147.5	146±30	137.5	120	159±43	177.4
165	92 ±18	89.6	152.5	142±25	131.9	130	157±34	159.3
175	81 ±12	86.7	157.5	164±32	126.8	140	125±25	144.7
			162.5	138±19	122.1	150	111±19	132.7
			167.5	113±16	117.6	160	115±16	122.7

$$r_R^{\text{exp}} = 0.468 \pm 0.010 \qquad r_R^{\text{the}} = 0.475 \qquad \chi^2 = 0.5$$

EFT are also in good agreement with data for total cross sections at higher energy [53, 54] which were likewise not included in the fitting procedure, as can be seen in Fig. 6.

The Λp and $\Sigma^+ p$ scattering lengths and effective ranges are listed in Table 4 together with the corresponding hypertriton binding energies (pre-

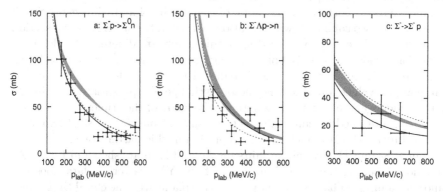

Fig. 6. As in Fig. 5, but now the experimental cross sections in (**a**),(**b**) are taken from [53] and in (**c**) from [54]

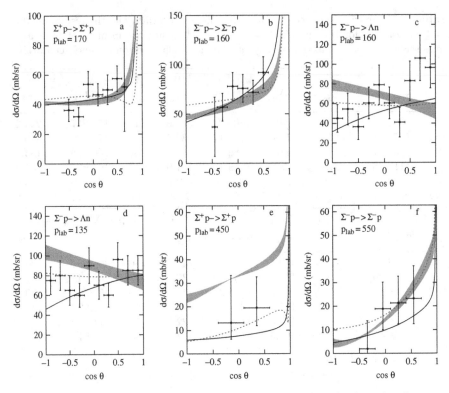

Fig. 7. Differential cross section $d\sigma/d\cos\theta$ as a function of $\cos\theta$, where θ is the c.m. scattering angle, at various values of p_{lab} (MeV/c). The experimental differential cross sections in (**a**),(**b**) are taken from [40], in (**c**),(**d**) from [41], in (**e**) from [55] and in (**f**) from [54]. Same description of curves as in Fig. 5

liminary results of YNN Faddeev calculations from [56]). The magnitudes of the Λp singlet and triplet scattering lengths obtained within chiral EFT are smaller than the corresponding values of the Jülich '04 and Nijmegen NSC97f models (last two columns), which is also reflected in the small Λp cross section near threshold, see Fig. 5a. But despite this significant difference the EFT interaction yields a correctly bound hypertriton too, see last row in Table 4. The singlet Σ^+p scattering length predicted by chiral EFT is about half as large as the values found for the meson-exchange YN potentials. Like in the latter models and other YN interactions, the value of the triplet Σ^+p scattering length obtained by chiral EFT is fairly small. Contrary to NSC97f, but as in the Jülich '04 YN model, there is repulsion in this partial wave.

Some S- and D-wave phase shifts for Λp and Σ^+p are shown in Fig. 8. As mentioned before, the limited accuracy of the YN scattering data does not allow for a unique phase shift analysis. This explains why the chiral EFT phase shifts are quite different from the phase shifts of the new meson-exchange YN

Table 4. The YN singlet and triplet scattering lengths and effective ranges (in fm) and the hypertriton binding energy, E_B (in MeV). The binding energies for the hypertriton (last row) [56] are calculated using the Idaho-N3LO NN potential [16]. The experimental value of the hypertriton binding energy is $-2.354(50)$ MeV [57]. We notice that the deuteron binding energy is -2.224 MeV

| Λ [MeV] | EFT '06 | | | | Jülich '04 | NSC97f [7] |
	550	600	650	700		
$a_s^{\Lambda p}$	-1.90	-1.91	-1.91	-1.91	-2.56	-2.51
$r_s^{\Lambda p}$	1.44	1.40	1.36	1.35	2.75	3.03
$a_t^{\Lambda p}$	-1.22	-1.23	-1.23	-1.23	-1.66	-1.75
$r_t^{\Lambda p}$	2.05	2.13	2.20	2.27	2.93	3.32
$a_s^{\Sigma^+ p}$	-2.24	-2.32	-2.36	-2.29	-4.71	-4.35
$r_s^{\Sigma^+ p}$	3.74	3.60	3.53	3.63	3.31	3.14
$a_t^{\Sigma^+ p}$	0.70	0.65	0.60	0.56	0.29	-0.25
$r_t^{\Sigma^+ p}$	-2.14	-2.78	-3.55	-4.36	-11.54	-25.35
$E_B(^3_\Lambda H)$	-2.35	-2.34	-2.34	-2.36	-2.27	-2.30

interaction of the Jülich group but also from all models presented in [7]. Indeed, the predictions of the various meson-exchange models also differ between each other in most of the partial waves. In both the Λp and $\Sigma^+ p$ $^1S_0$ and $^3S_1$ partial waves, the LO chiral EFT phase shifts are much larger at higher energies than the phases of the meson-exchange models. But this is not surprising. First we want to remind the reader that the empirical data YN considered in the fitting procedure are at lower energies. Second, also for the NN interaction in leading order these partial waves were much larger than the Nijmegen phase shift analysis, see [34]. It is expected that this problem for the YN interaction can be solved by the derivative contact terms in a NLO calculation, just like in the NN case. It is interesting to see that the $^3S_1$ $\Sigma^+ p$ phase shift is repulsive in chiral EFT as well as in the new Jülich meson-exchange model, but attractive in the Nijmegen NSC97f model. This has consequences for the $\Sigma^+ p$ differential cross section because, depending on the sign, the interference of the hadronic amplitude with the Coulomb amplitude differs, cf. Fig. 7. Unfortunately, the limited accuracy of the available $\Sigma^+ p$ data does not allow to discriminate between these two scenarios.

Results for P-wave phase shifts can be found in [7, 9, 26]. Here we just want to remark that in case of LO chiral EFT the P-waves are the result of pseudoscalar meson exchange alone, since we only have contact terms in the S-waves in that order. Also, contrary to conventional meson-exchange models, in LO chiral EFT there are no spin singlet to spin triplet transitions, because of the potential form in (16) and (30). Although the $^3D_1$ Λp phase shift near

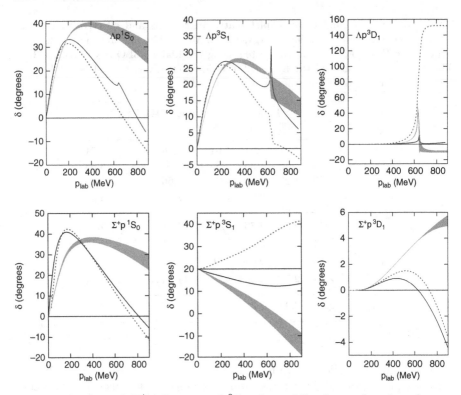

Fig. 8. The Λp and Σ^+p S-wave and $^3D_1$ phase shifts δ as a function of p_{lab}. Since the phases of the Jülich '04 model are calculated in the isospin basis, its ΣN threshold does not exactly coincide with the others. Same description of curves as in Fig. 5

the ΣN threshold rises quickly for our YN interactions, cf. Fig. 8, it does not go through 90 degrees in both cases – unlike the Nijmegen model NSC97f [7]. The opening of the ΣN channel is also clearly seen in the $^3S_1$ Λp partial wave for all considered interactions, but again there are significant differences in the concrete behavior.

Very recently the chiral EFT model has been employed in Faddeev-type investigations of the four-body systems $^4_\Lambda$H and $^4_\Lambda$He [58]. The binding energies of these hypernuclei are especially interesting predictions. It has been very difficult in the past to describe their charge symmetry breaking (CSB) and the splitting of the 0^+ ground and 1^+ excited state at the same time [59]. In Table 5, we show the differences of the binding energies of the core nucleus and the hypernucleus, the Λ separation energies, since these are only mildly dependent on the NN interaction model used for the calculations [59]. We compare the Λ separation energies based on chiral EFT and the two considered meson-exchange YN interactions to the experimental numbers. It is seen that the separation energies for the excited states are somewhat dependent on the

Table 5. Λ separation energies of the 0^+ ($E_{\mathrm{sep}}(0^+)$) and 1^+ ($E_{\mathrm{sep}}(1^+)$) states and their difference ΔE_{sep} for ${}^4_\Lambda\mathrm{H}$ and the difference of the separation energies for the mirror hypernuclei ${}^4_\Lambda\mathrm{He}$ and ${}^4_\Lambda\mathrm{H}$ (CSB-0^+ and CSB-1^+). Results for the chiral EFT YN interaction for various cut–offs Λ are compared to predictions for the Jülich '04 and Nijmegen NSC97f meson-exchange models and the experimental values [57]

Λ [MeV]	EFT '06				Jülich '04	NSC97f	Expt.
	500	550	650	700			
$E_{\mathrm{sep}}(0^+)$ [MeV]	2.63	2.46	2.36	2.38	1.87	1.60	2.04
$E_{\mathrm{sep}}(1^+)$ [MeV]	1.85	1.51	1.23	1.04	2.34	0.54	1.00
ΔE_{sep} [MeV]	0.78	0.95	1.13	1.34	-0.48	0.99	1.04
CSB-0^+ [MeV]	0.01	0.02	0.02	0.03	-0.01	0.10	0.35
CSB-1^+ [MeV]	-0.01	-0.01	-0.01	-0.01	—	-0.01	0.24

cut–off value chosen. Certainly, contributions from higher order will be sizable for these observables. However, within these uncertainties, the results agree remarkably well with the experimental separation energies, which is somewhat less the case for the meson-exchange potentials. The CSB of the separation energies is not well described by all of the interactions. The Nijmegen model NSC97f includes explicit CSB in the potential, which induces a sizable but too small effect on the separation energies. It will be interesting to study this observable in NLO of the chiral interaction, where first explicitly CSB terms contribute.

6 Summary and Outlook

In this review we presented results based on two different approaches to the YN interaction, namely on the traditional meson-exchange picture and on chiral effective field theory.

As far as meson-exchange models of the YN interaction are concerned we focussed on the recent model of the Jülich group, whose main new feature is that the contributions both in the scalar-isoscalar (σ) and the vector-isovector (ϱ) channels are constrained by a microscopic model of correlated $\pi\pi$ and $K\overline{K}$ exchange. Besides those contributions from correlated $\pi\pi$ and $K\overline{K}$ exchange this model incorporates also the standard one-boson exchanges of the lowest pseudoscalar and vector meson multiplets with coupling constants fixed by SU(6) symmetry relations. Thus, the long- and intermediate-range part of this YN interaction model is completely determined – either by SU(6) constraints or by correlated $\pi\pi$ and $K\overline{K}$ exchange.

The YN interaction derived within chiral EFT is based on a modified Weinberg power counting, analogous to the NN force in [15]. The symmetries of QCD are explicitly incorporated. Also here it is assumed that the interactions in the various YN channels are related via $SU(3)_f$ symmetry. However, since we have done our study in leading order, in which the NN interaction can not be described well, we do not connect the present YN interaction with the NN sector, but focus on the YN system only.

To be specific, the leading-order potential consists of two pieces: firstly, the longer-ranged one-pseudoscalar-meson exchanges, related via $SU(3)_f$ symmetry in the well-known way and secondly, the shorter ranged four-baryon contact term without derivatives. The latter contains five independent low-energy constants that need to be determined from the empirical data. We fixed those five free parameters by fitting to 35 low-energy YN scattering data. The reaction amplitude is obtained by solving a regularized Lippmann-Schwinger equation for the chiral EFT interaction. The regularization is done by multiplying the strong potential with an exponential regulator function where we used a cut-off in the range between 550 and 700 MeV.

The meson-exchange picture has been already applied successfully to the YN system in the past by many authors. Thus, it is not surprising that a good reproduction of the data could be achieved within this approach. But it is rather reassuring to see that also chiral effective field theory works remarkably well for the YN interaction, in particular since we have, so far, restricted ourselves to lowest order only. Indeed, we could obtain a rather good description of the empirical data, as is reflected in the total χ^2 which is the range between 28.3 and 34.6 for a cut-off in the range between 550 and 700 MeV. In addition low-energy differential cross sections and higher energy cross sections, that were not included in the fitting procedure, were predicted quite well.

In a first application to few-baryon systems involving strangeness we found that the chiral EFT yields a correctly bound hypertriton [56]. We did not explicitly include the hypertriton binding energy in the fitting procedure, but we have fixed the relative strength of the ΛN singlet and triplet S-waves in such a way that a bound hypertriton could be obtained. It is interesting to note that a Λp singlet scattering length of -1.9 fm leads to the correct binding energy. Meson-exchange YN models that yield comparable results for the hypertriton binding energy predict here singlet scattering lengths that are typically in the order of -2.5 fm.

In conclusion, our results strongly suggest that the chiral effective field theory scheme, applied in [15] to the NN interaction, also works well for the YN interaction. In the future it will be interesting to study the convergence of the chiral EFT for the YN interaction by doing NLO and NNLO calculations. In particular a combined NN and YN study in chiral EFT, starting with a NLO calculation, needs to be performed. Also an $SU(3)$ extension to the hyperon-hyperon (YY) sector is of interest. In this case only one additional low-energy constant arises within the EFT approach in LO. This constant could be fixed by available data on the reaction $\Xi^- p \to \Lambda\Lambda$ [60], say, and then predictions can be made for all reaction channels in the strangeness -2 sector. In particular, one would then be able to obtain an estimate for the $\Lambda\Lambda$ interaction, whose strength is rather crucial for the existence of doubly strange hypernuclei. With regard to the interactions presented in this paper it will be interesting to see their performance when employed in further calculations of strange few-baryon systems as well as in hypernuclei. For example, preliminary results for the four-body hypernuclei $^4_\Lambda$H and $^4_\Lambda$He show that the chiral EFT predicts reasonable Λ separation energies for $^4_\Lambda$H, though the charge dependence of the Λ separation energies is not reproduced (as expected at lowest order).

References

1. R. Machleidt, K. Holinde, and C. Elster: Phys. Rep. **149**, 1 (1987)
2. R. Machleidt and I. Slaus: J. Phys. **G27**, R69 (2001)
3. P. M. M. Maessen, T. A. Rijken, and J. J. de Swart: Phys. Rev. C **40**, 2226 (1989)
4. B. Holzenkamp, K. Holinde, and J. Speth: Nucl. Phys. A **500**, 485 (1989)
5. A. Reuber, K. Holinde, and J. Speth: Nucl. Phys. A **570**, 543 (1994)
6. K. Tominaga et al: Nucl. Phys. A **642**, 483 (1998)
7. T. A. Rijken, V. G. J. Stoks, and Y. Yamamoto: Phys. Rev. C **59**, 21 (1999)
8. K. Tominaga and T. Ueda: Nucl. Phys. A **693**, 731 (2001)
9. J. Haidenbauer and U. -G. Meißner: Phys. Rev. C **72**, 044005 (2005)
10. T. A. Rijken and Y. Yamamoto: Phys. Rev. C **73**, 044008 (2006)
11. E. Klempt, F. Bradamante, A. Martin, and J. -M. Richard: Phys. Rep. **368**, 119 (2002)
12. P. F. Bedaque and U. van Kolck: Annu. Rev. Nucl. Part. Sci. **52**, 339 (2002)
13. D. Kaplan: Lectures delivered at the 17th National Nuclear Physics Summer School 2005, Berkeley, CA, June 6–17, 2005; nucl-th/0510023
14. E. Epelbaum: Prog. Nucl. Part. Phys. **57**, 654 (2006)
15. E. Epelbaum, W. Glöckle, and U. -G. Meißner: Nucl. Phys. A **747**, 362 (2005)
16. D. R. Entem and R. Machleidt: Phys. Rev. C **68**, 041001 (2003)
17. S. Weinberg: Phys. Lett. B **251**, 288 (1990)
18. S. Weinberg: Nucl. Phys. B **363**, 3 (1991)
19. E. Epelbaoum, W. Glöckle, and U. -G. Meißner: Nucl. Phys. A **637**, 107 (1998)
20. M. J. Savage and M. B. Wise: Phys. Rev. D **53**, 349 (1996)
21. H. W. Hammer: Nucl. Phys. A **705**, 173 (2002)
22. C. L. Korpa, A. E. L. Dieperink, and R. G. E. Timmermans: Phys. Rev. C **65**, 015208 (2001)
23. D. B. Kaplan, M. J. Savage, and M. B. Wise: Nucl. Phys. B **534**, 329 (1998)
24. S. R. Beane, P. F. Bedaque, A. Parreño, and M. J. Savage: Nucl. Phys. A **747**, 55 (2005)
25. S. R. Beane et al: hep-lat/0612026
26. H. Polinder, J. Haidenbauer, and U. -G. Meißner: Nucl. Phys. A **779**, 244 (2006)
27. C. M. Vincent and S. C. Phatak: Phys. Rev. C **10**, 391 (1974)
28. M. Walzl, U. -G. Meißner, and E. Epelbaum: Nucl. Phys. A **693**, 663 (2001)
29. J. D. Bjorken and S. D. Drell: *Relativistic Quantum Fields* (McGraw-Hill Inc., New York, 1965). We follow the conventions of this reference
30. T. -P. Cheng and L. -F. Li: *Gauge theory of elementary particle physics* (Oxford University Press, Oxford, 1984)

31. T. A. Rijken, R. A. M. Klomp, and J. J. de Swart: in *A Gift of Prophecy, Essays in Celebration of the Life of Robert Eugene Marshak*, ed by E. C. G. Sudarshan (World Scientific, Singapore, 1995)
32. J. J. de Swart: Rev. Mod. Phys. **35**, 916 (1963)
33. C. B. Dover and H. Feshbach: Ann. Phys. **198**, 321 (1990)
34. E. Epelbaum, W. Glöckle, and U. -G. Meißner: Nucl. Phys. A **671**, 295 (2000)
35. U. -G. Meißner: Rep. Prog. Phys. **56**, 903 (1993)
36. E. Epelbaum et al: Eur. Phys. J. A **15**, 543 (2002)
37. E. Epelbaum, W. Glöckle, and U. -G. Meißner: Eur. Phys. J. A **19**, 401 (2004)
38. B. Sechi-Zorn, B. Kehoe, J. Twitty, and R. A. Burnstein: Phys. Rev. **175**, 1735 (1968)
39. G. Alexander et al: Phys. Rev. **173**, 1452 (1968)
40. F. Eisele, H. Filthuth, W. Fölisch, V. Hepp, and G. Zech: Phys. Lett **37B**, 204 (1971)
41. R. Engelmann, H. Filthuth, V. Hepp, and E. Kluge: Phys. Lett. **21**, 587 (1966)
42. J. J. de Swart and C. Dullemond: Ann. Phys. **19**, 485 (1962)
43. M. M. Nagels, T. A. Rijken, and J. J. de Swart: Phys. Rev. D **15**, 2547 (1977)
44. E. Klempt: *Glueballs, Hybrids, Pentaquarks : Introduction to Hadron Spectroscopy and Review of Selected Topics*, Lectures at the 18th Annual Hampton University Graduate Studies, Jefferson Lab, Newport News, VA, June 2–20, 2003; hep-ph/0404270
45. Y. Kalashnikova, A. E. Kudryavtsev, A. V. Nefediev, J. Haidenbauer, and C. Hanhart: Phys. Rev. C **73**, 045203 (2006)
46. A. Reuber, K. Holinde, H. -C. Kim, and J. Speth: Nucl. Phys. A **608**, 243 (1996)
47. G. Janssen, B. Pierce, K. Holinde, and J. Speth: Phys. Rev. D **52**, 2690 (1995)
48. D. Lohse, J. Durso, K. Holinde, and J. Speth: Nucl. Phys. A **516**, 513 (1990)
49. G. Janssen, K. Holinde, and J. Speth: Phys. Rev. C **54**, 2218 (1996)
50. P. La France and P. Winternitz: J. Physique **41**, 1391 (1980)
51. J. A. Kadyk, G. Alexander, J. H. Chan, P. Gaposchkin, and G. H. Trilling: Nucl. Phys. B **27**, 13 (1971)
52. J. M. Hauptman, J. A. Kadyk, and G. H. Trilling: Nucl. Phys. B **125**, 29 (1977)
53. D. Stephen: PhD thesis, University of Massachusetts, 1975, unpublished
54. Y. Kondo et al: Nucl. Phys. A **676**, 371 (2000)
55. J. K. Ahn et al: Nucl. Phys. A **648**, 263 (1999)
56. A. Nogga, J. Haidenbauer, H. Polinder, and U. -G. Meißner: in preparation
57. B. F. Gibson and E. V. Hungerford: Phys. Rep. **257**, 349 (1995)
58. A. Nogga: *Application of chiral nuclear forces to light nuclei*, 5th Int. Workshop on Chiral Dynamics, Theory and Experiment, Durham/Chapel Hill, NC, Sept. 18–22, 2006; nucl-th/0611081
59. A. Nogga, H. Kamada, and W. Glöckle: Phys. Rev. Lett. **88**, 172501 (2002)
60. J. Ahn et al: Phys. Lett. B **633**, 214 (2006)

Weak Decays of Hypernuclei

Assumpta Parreño

Dept. d'Estructura i Constituents de la Matèria, Universitat de Barcelona,
Barcelona, 08028-Spain
aparreno@ub.edu

Abstract. I present the main theoretical attempts carried out during the last few years to understand the weak hyperon-nucleon interaction through the comparison with hypernuclear decay observables, as well as the present perspectives and research avenues in the field.

The manuscript is organized as follow. The field of hypernuclear decay is presented in the Introduction. In Sect. 2 I discuss the possible modes for a hyperon to decay, the mesonic channel ($\Lambda \to \pi N$) and the non-mesonic channel ($\Lambda N \to NN$), which constitutes the dominant mode when the decay takes place in the medium. In Sect. 3 I concentrate in two different methods to evaluate the mesonic decay rate in a finite nucleus, the Propagator Method and the Wave Function Method. In Sect. 4 I present the finite nucleus calculation of the non-mesonic decay rate, again using two methods, the Polarization Propagator Method supplied by a Local Density Approximation, and a direct Finite Nucleus calculation. In this same Section, I discuss how to calculate the partial and total decay rates, and present the various models existing in the literature to describe the weak $|\Delta S| = 1$ $\Lambda N \to NN$ transition potential, with special emphasis in the One-Meson-Exchange model. Sections 5 and 6 are devoted to the study of the role played by the strong interaction in hypernuclear decay, paying special attention to the effect of final state interactions of the nucleons emitted after the weak decay with the residual medium. The theoretical framework to evaluate the asymmetry in the distribution of protons coming from the decay of polarized hypernuclei is approached in Sect. 7, with the corresponding comparison with present experimental data. Recent attempts to obtain a model independent description of the weak decay process are tackled in Sect. 8. And finally, Sect. 9 presents the summary and some issues that, in my opinion, are especially worthy to address in the future.

1 Introduction

Hypernuclei are bound systems of nucleons where we have added one or more impurities, strange baryons (hyperons), which besides the basic components of nucleons, the up and down quarks, contain the strange quark. These systems provide a wonderful framework to learn about nuclear and particle physics.

A. Parreño: *Weak Decays of Hypernuclei*, Lect. Notes Phys. **724**, 141–189 (2007)
DOI 10.1007/978-3-540-72039-3_5 © Springer-Verlag Berlin Heidelberg 2007

The distinguishability of the hyperon from nucleons makes it a privileged probe to explore states deep inside the nucleus, extending our knowledge of conventional to flavored nuclear physics. Being not affected by the Pauli exclusion principle, the hyperon in a single-hyperon hypernuclei can sit at the center of a nucleus. This fact has suggested changes in the nuclear dynamics based on the interaction of the hyperon with the surrounding nucleons, which could produce changes in the size and shape of nuclei, making the nucleus shrink and changing its density. It has been speculated for instance, that the glue–like role of the Λ hyperon can facilitate the existence of neutron-rich hypernuclei, being a more suitable framework to study matter with extreme neutron to proton ratios as compared to ordinary nucleons.

Moreover, the study of hypernuclei allows us to get insight into fundamental interactions where particle reactions are not accessible, as it is the case of the hyperon-nucleon and hyperon-hyperon interactions. Quantitative information on these interactions is only possible through the study of the production and decay of hypernuclei, since their study in free space is hindered by the instability of hyperons against the weak interaction. With a typical lifetime of the order of 10^{-10} s, except the Σ^0 which decays electromagnetically $\Sigma^0 \rightarrow \Lambda + \gamma$ much faster, hyperons decay through reactions which do not conserve strangeness, isospin nor parity. The Λ particle is the lightest among the hyperons and decays into protons, $\Lambda \rightarrow p\pi^-$ (64%), and into neutrons, $\Lambda \rightarrow n\pi^0$ (36%). These reactions, called mesonic decay modes, are strongly suppressed when the hyperon sits in the nuclear medium, due to the low momentum of the outgoing nucleon (around $100\,\mathrm{MeV/c}$) compared to the typical Fermi momentum in nuclear matter ($\sim 270\,\mathrm{MeV/c}$). But what experiments see is that hypernuclei decay, leaving not only one nucleon behind but two, three or more nucleons, together with residual non-strange bound fragments. This is due to the appearance of new decay channels induced by the presence of the surrounding nucleons, the non-mesonic modes, which can be induced by one or more nucleons, $\Lambda N \rightarrow NN$, $\Lambda NN \rightarrow NNN$, etc. Hypernuclear decay offers then the possibility to study the weak baryon-baryon interaction using the change of strangeness as a signature. This is in contrast to what happens in the $\Delta S = 0$ sector, where the weak nucleon-nucleon (NN) signal, which has a parity-conserving (PC) component, is masked by the much larger NN strong signal, which also conserves parity[1].

While the mesonic decay mode of the Λ hyperon gets (Pauli) blocked by the presence of the nucleons in the medium, the *non-mesonic* mode, in particular the one-nucleon induced reaction $\Lambda N \rightarrow nN$, becomes the predominant

---

[1] Taking the strong coupling constant of order $\alpha_S \sim 1$ in the energy range around $1\,\mathrm{GeV}$, the weak coupling constant lies in the range of $\alpha_W \sim 10^{-6} - 10^{-7}$. This value can be obtained by comparing the lifetimes of the Δ^+ and Σ^+ baryons, since the strength of the interaction is related to the inverse square of the coupling. These particles decay into the same final state products, a proton and a neutral pion, but through different mechanisms, the strong interaction the former and the weak $\Delta S = -1$ interaction the later, of much smaller strength.

decay channel for medium to heavy hypernuclei. For many years, experimental facilities have successfully created hypernuclei and studied their decay. The theoretical study of such decays requires a quite involved calculation, since the elementary reaction takes place in the medium and nuclear structure details as well as the propagation of the outgoing nucleons through the residual nuclear medium have to be taken into account. In spite of this, different sophisticated models have been developed to accommodate the available data for the decay observables. While the total decay rate is fairly reproduced by most of the theoretical models, the situation has been more controversial during decades for the rest of observables, namely, the ratio between the neutron-induced channel and the proton-induced one, Γ_n/Γ_p, and the asymmetry in the angular distribution of final protons. Although the data was very rough years ago, making difficult to extract reliable information on the underlying hyperon-nucleon weak dynamics, present sophisticated experiments are very accurate, and provide us with a much cleaner information.

In these lectures I present the main theoretical attempts performed during the last few years to explain the available experimental data, as well as the present perspectives and lines of research. More detailed and diverse information can be found in any of the good reviews on the subject presented in the literature. See for instance [1, 2] and the *Enrico Fermi school* proceedings, [3], for the latest updated publications, and references therein.

2 Hypernuclear Decay Modes

In 1952, the Polish scientists M. Danysz and J. Pniewski observed the first hypernuclear decay event in a photographic emulsion exposed to cosmic rays at around 26 km above the ground [4]. This event meant the birth of hypernuclear physics. The observation is depicted in Fig. 1, where we can see an incoming high energy proton from the top, which collides with one of the nuclei in the emulsion breaking it into different nuclear fragments (star shape produced in point "A" in the picture). These fragments, after a short path, eventually stop in the emulsion, but one, after traveling a path length of 90 μm, disintegrates into three more particles (point "B" in the figure), revealing the presence of an unstable particle stuck among the nucleons. This event can be interpreted as the weak decay of a light fragment containing a hyperon. Since then, the field has seen a lot of activity, from the theoretical and experimental side. Experimentally, hadronic reactions for the production of strange bound systems include: i) strangeness-exchange reactions, performed at CERN (Switzerland), BNL (USA), KEK (Japan) and DAΦNE (Italy), as the $N(K^-, \pi^-)\Lambda$ one, where the strangeness transfers from the initial to the final state, ii) strangeness associated production reactions, as the $n(\pi^+, K^+)\Lambda$ one at BNL and KEK, where a $s\bar{s}$ pair is created in the final state, and iii) electro production mechanisms, as the $p(e, e'K^+\Lambda)$ at TJNAF (USA) and GSI (Germany), which also produce $s\bar{s}$ final pairs. The hypernucleus is then

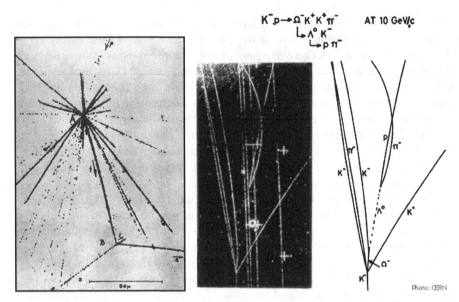

Fig. 1. Left panel: first hypernuclear decay event [4]. **Right panel**: A photograph from the 2m hydrogen filled bubble chamber at CERN. Five K⁻ mesons enter the chamber from bottom and one of them interacts with a proton in the chamber. In the interaction (reconstruction of the interaction to the right) a very rare Ω^- particle, containing three strange quarks, is produced. The magnetic field bends the charged particles. The Ω^- decays after about 10 cm in the chamber into one Λ particle (neutral) and one K⁻. The white crosses (+) are marks fixed to the chamber that help the three dimensional reconstruction. The picture and text have been taken from http://nobelprize.org/

produced in an excited state and rapidly reaches the ground state by electromagnetic gamma or particle emission. Once stable against strong decay modes, hypernuclei decay weakly through processes that do not conserve parity, strangeness nor isospin. While the mesonic decay mode of the Λ hyperon gets blocked by the presence of the nucleons in the medium, it is this same medium which promotes the non-mesonic mode, according to which the Λ interacts with one (or more) of the surrounding nucleons. A simple single-particle shell-model is appropriate to picture the distribution of particles in a hypernucleus. One can think of two different wells for protons and neutrons, where nucleons occupy the corresponding energy levels according to their quantum numbers and the Pauli exclusion principle, filling the shells up to the Fermi level. The Λ hyperon, with a (uds) quark content, is the lightest among the hyperons, with a mass of $m_\Lambda = 1115.684 \pm 0.006\,\mathrm{MeV/c^2}$, approximately a 20% larger than the nucleon mass, zero total charge and isospin, positive parity and total angular momentum $j = \frac{1}{2}$. Since it carries a new quantum number, strangeness $(S = -1)$, it can occupy states already filled by nucleons, and therefore, explore deep bound states in nuclear systems.

The free Λ particle has a lifetime of $\tau_\Lambda \equiv \hbar/\Gamma_\Lambda = 2.632 \times 10^{-10}$ s, and as already mentioned, its main decay modes are the mesonic $\Lambda \to n \pi^0$ and $\Lambda \to p \pi^-$ channels, through diagrams as the one depicted in Fig. 2. There are other decay modes though, less relevant, as the semi-leptonic and radiative decay modes, with branching ratios much smaller,

$$
\Lambda \to \begin{cases} n\gamma & (\text{B.R.} = 1.75 \times 10^{-3}); \\ p\pi^-\gamma & (\text{B.R.} = 8.4 \times 10^{-4}); \end{cases} \quad \begin{matrix} pe^-\bar{\nu}_e \ (\text{B.R.} = 8.32 \times 10^{-4}) \\ p\mu^-\bar{\nu}_\mu \ (\text{B.R.} = 1.57 \times 10^{-4}) \end{matrix} .
$$

In the medium, the difference between the Λ and N masses facilitates the emission of two fast nucleons from the non-mesonic decay which can then access to non occupied states in a shell-model picture. As can be seen in Fig. 3, the analysis of hypernuclear lifetimes as a function of the mass number, A, shows that the mesonic decay (MD) mode gets blocked as A increases, while the non-mesonic decay (NMD) increases up to a saturation value of the order of the free Λ decay, reflecting the short-range nature of the weak $\Delta S = 1$ baryon-baryon interaction.

2.1 Weak Decay and the $\Delta I = 1/2$ Rule

The free decay of the Λ hyperon leaves a nucleon and a pion in the final state which can be coupled to isospin $1/2$ or $3/2$. Since the Λ is an isoscalar, this means that the weak transition can in principle carry $1/2$ and $3/2$ units of isospin. A simple Clebsch-Gordan coefficients analysis tells us that if we assume $\Delta I = 1/2$ transitions, the ratio between the decay into π^- over the decay into π^0 is roughly given by:

$$
\frac{\Gamma^{free}_{\Lambda \to \pi^- p}}{\Gamma^{free}_{\Lambda \to \pi^0 n}} \sim \frac{|\langle \pi^- p \mid T_{1/2,-1/2} \mid \Lambda \rangle|^2}{|\langle \pi^0 n \mid T_{1/2,-1/2} \mid \Lambda \rangle|^2} = \frac{|\sqrt{2/3}|^2}{|\sqrt{1/3}|^2} = 2 ,
\tag{1}
$$

while, $\Delta I = 3/2$ would give us:

$$
\frac{\Gamma^{free}_{\Lambda \to \pi^- p}}{\Gamma^{free}_{\Lambda \to \pi^0 n}} \sim \frac{|\langle \pi^- p \mid T_{3/2,-1/2} \mid \Lambda \rangle|^2}{|\langle \pi^0 n \mid T_{3/2,-1/2} \mid \Lambda \rangle|^2} = \frac{|\sqrt{1/3}|^2}{|\sqrt{2/3}|^2} = \frac{1}{2} .
\tag{2}
$$

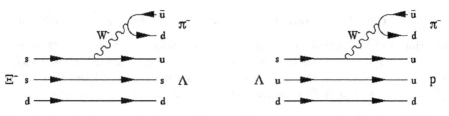

Fig. 2. Quark diagrams for the weak decays of hyperons through coupling of a W^- boson to a s to u quark line. **Left:** decay of the Ξ^- baryon. **Right:** decay of the Λ hyperon

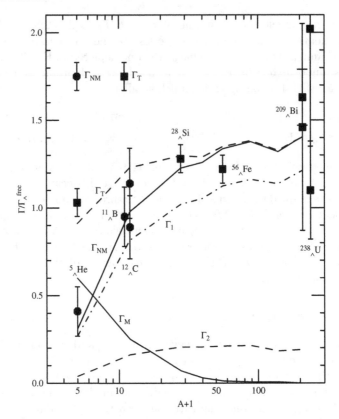

Fig. 3. Weak decay rates as function of the total number of particles. The lines are theoretical estimations obtained with a Polarization Propagator Method supplemented by a Local Density Approximation (see text). The *upper dashed line* stands for the total decay rate, Γ_T, the decreasing solid line represents the mesonic decay mode, Γ_M, while the increasing solid line represents the total non-mesonic decay rate, Γ_{NM}, which corresponds to the sum of the one-nucleon induced decay (*dot-dashed*), Γ_1, and the two-nucleon induced mode (*lower dashed line*), Γ_2. Experimental values of the total and non-mesonic decay rates are given by the square and circle marks respectively (see [5] and references therein)

Experimentally, this ratio has been measured[2] to be $\dfrac{\Gamma_{\Lambda \to \pi^- p}^{free}}{\Gamma_{\Lambda \to \pi^0 n}^{free}} = 1.78$, indicating a clear dominance of $\Delta I = 1/2$ transitions over $\Delta I = 3/2$ ones. Even though our previous theoretical estimation, assuming same phase space for both processes and the absence of final state interactions, looks very simplistic, it is a clear indication that at this level, $\Delta I = 3/2$ transitions are

---

[2] The experimental observation of this value involves also the determination of some polarization observables for the nonleptonic decay of hyperons, besides the total mesonic decay rate. See Sect. XII-6 of [6].

very much suppressed. This experimentally verified property of the weak interaction is called the $\Delta I = 1/2$ *rule*, and its dynamical origin is not yet understood on theoretical grounds. Moreover, related fundamental questions arise that still have to be answered. Is this a universal feature of the weak $\Delta S \neq 0$ interactions? Does this rule hold also in the decay of hyperons in the medium?

The mesonic width is also very sensitive to the Q value of the process, $Q \approx m_\Lambda - m_N - m_\pi \approx 37\,\text{MeV}$, or in other words, to the center-of-mass momentum, $|\boldsymbol{p}| \approx 100\,\text{MeV}/c$. This implies a large sensitivity of the available phase space to the mass of the final light particle (pion) and to the Λ and final nucleon binding energies. The difference between these binding energies reaches the value $B_\Lambda - B_N \sim -27 + 8\,\text{MeV}$ in heavy hypernuclei, producing an even smaller Q value in the medium with respect to the free space decay, and therefore, an even smaller value of the nucleon momentum. In addition, a small effect that contributes to the reduction of the MD rate in the medium is the absorption of the pion by the nucleons in the medium. All these considerations make the mesonic decay mode strictly forbidden in infinite nuclear matter. But experiments deal with finite nuclear systems, and in these systems the MD can proceed due to different reasons.

Besides the fact that the local Fermi momentum is smaller at the nuclear surface, allowing for more available states for the final nucleons as compared to the nuclear matter calculation, the hyperon in the medium has no zero momentum, as opposed to the decay in free space, but some momentum distribution due to its spatial confinement. This allows the final nucleon to carry larger momentum and to overcome the Pauli blocking. Moreover, the MD mode shows a strong sensitivity to the pion nucleus optical potential. The total mesonic decay width is significantly enhanced due to the pion attraction by the medium, U_π, which comes basically from the attractive p-wave part of the potential. Therefore, for a fixed momentum $\boldsymbol{q}$, the pion has an energy smaller than the free one, and due to energy conservation, $E_\Lambda = E_N + \sqrt{(q^2 + m_\pi^2)} + U_\pi$, the final nucleon has more chances to go above the Fermi surface. This increase can be of one or two orders of magnitude for heavy hypernuclei [1, 7]. The effect is smaller for light and medium hypernuclei (a factor of 2 for A = 16). This gigantic increase on the rate becomes more moderate when one looks also to exclusive reactions to a final closed shell nucleus, which select basically the repulsive s-wave part of the pion-nucleus optical potential. One has to perform therefore a simultaneous study of inclusive and exclusive reactions to use the MD channel as a reliable complementary source of information on the pion-nucleus interactions.

Nevertheless, at the end, the mesonic decay mode decreases as the mass number increases, being overshadowed by the more prominent non-mesonic decay mode. This mode can be easily understood in terms of a one-meson-exchange (OME) mechanism, where the meson emitted at the weak vertex is being absorbed by one of the nucleons in the medium. Traditionally, after the first phenomenological approaches [8] and in analogy to the strong nucleon-nucleon interaction, the $|\Delta S| = 1$ ΛN interaction was studied on the basis of OME models [9, 10, 11], or through models which combined this meson-exchange picture with effective quark Hamiltonians [12] to describe different energy ranges. The elementary weak two-body process has to be folded by the strong interaction, obviously present in any physical process involving baryons. Regarding the initial hypernucleus, a reasonable approach consists in assuming a shell-model. Within this picture, deexcitation modes will rapidly place the Λ particle in the lowest energy level, $1s_{1/2}$, before the weak decay occurs. The wave functions for the hyperon and the nucleon are taken such that the binding energy of the hypernucleus and the charge form-factor of the nuclear core are reproduced. After the weak decay takes place, the two primary nucleons propagate in the medium, interacting strongly with the other nucleons, changing their momentum, direction and charge. Present day calculations, with all these ingredients carefully taken into account, have been very successful in reproducing the experimental total and partial decay rates for the most commonly studied hypernuclei [13, 14].

3 Finite Nucleus Calculation of the Mesonic Decay Rate

The decay of finite systems has been approached by two different methods, the Polarization Propagator Method (PPM) and the Wave Function Method (WFM). The first one relies on a many-body description of the hyperon self-energy in nuclear matter, while the finite nuclei calculation is performed through the Local Density Approximation (LDA). The second one is a direct finite nucleus calculation which uses appropriate shell model nuclear and hypernuclear wave functions (at hadronic and quark level). In both methods, when the pion wave function has to be implemented (when evaluating the mesonic decay width for instance), one uses the appropriate wave function generated by pion-nucleus potentials.

3.1 The Propagator Method. Mesonic Decay Width in Free Space

The starting point for the derivation is the weak ΛNπ Lagrangian,

$$\mathcal{L}_{\Lambda N \pi}^{W} = -i \, G_F m_\pi^2 \, \overline{\Psi}_N \left(A + B \, \gamma_5 \right) \boldsymbol{\tau} \cdot \boldsymbol{\phi}_\pi \Psi_\Lambda \begin{pmatrix} 0 \\ 1 \end{pmatrix} + \text{h.c.} \qquad (3)$$

where Ψ_N and Ψ_Λ are the free baryon fields of positive energy, ϕ_π the pion field, $G_F = 2.21 \times 10^{-7}/m_\pi^2$ the weak Fermi constant, and the constants A and B, are adjusted in order to reproduce the Parity-Violating (PV) and the Parity-Conserving (PC) strengths of the free width, respectively. Note that one could have used a derivative coupling for the pion to the baryons, instead of a pseudoscalar coupling, more appropriate from a chiral point of view, but the result is the same when one is dealing with positive energy states [16]. With respect to the isospin, theoretically we know how to handle processes where isospin is conserved through the vertex. This is not the case of the weak transition. In order to use isospin formalism, and at the same time to account for the $\Delta I = 1/2$ rule, one dresses the Λ hyperon, which has isospin zero, with $1/2$ units of isospin. To account for the zero charge of the Λ particle, we choose the projection to be $-1/2$. This is achieved by coupling an isospurion, $\binom{0}{1}$, to the Λ field.

The Λ width is directly related to the imaginary part of the Λ self-energy (Σ) diagram depicted in Fig. 4 under the label "free", as $\Gamma_\Lambda = -2\,\mathrm{Im}\Sigma$, and it is represented by the cut on the intermediate states. This cut puts the intermediate particles on shell, giving rise to the $\Lambda \to \pi N$ process. Using standard Feynman rules the self-energy can be written as:

$$-i\Sigma = 3(G_F m_\pi^2)^2 \int \frac{d^4 q}{(2\pi)^4}\, G(k-q)\, D(q) \left(S^2 + \frac{P^2}{m_\pi^2} q^2 \right) , \qquad (4)$$

where G and D stand for the free nucleon and pion propagators:

$$G(k-q) = \frac{1}{(k^0 - q^0) - E(\boldsymbol{k} - \boldsymbol{q}) + i\epsilon}$$

$$D(q) = \frac{1}{(q^0)^2 - \boldsymbol{q}^2 - m_\pi^2 + i\epsilon} , \qquad (5)$$

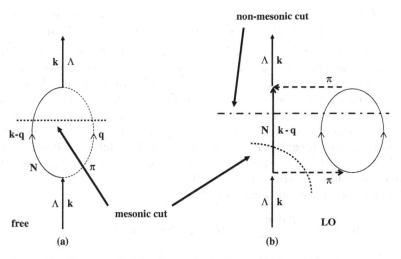

Fig. 4. Free Λ self-energy and leading order in the expansion of the pion propagator. To get the mesonic decay widths one has to perform the cuts depicted by the *dotted line* in each figure, where the nucleon and pion are placed on-shell, while the non-mesonic cut corresponds to the *dot-dashed line*, where the intermediate nucleons are put on shell

and the constants S and P are related to the PV and PC amplitudes of (3) by:

$$S = A, \quad P = \frac{m_\pi B}{2 \, m_N} .$$ (6)

After performing the q^0–integration, we can write the well known result for the (non-relativistic) free Λ width [15] as:

$$
\begin{aligned}
\Gamma_\Lambda^{\text{free}} &= 3(G_F m_\pi^2)^2 \\
&\times \int \frac{d^3q}{(2\pi)^3} \frac{1}{2\,\omega(q)} 2\pi \, \delta[\, E_\Lambda - \omega(q) - E_N(\boldsymbol{k} - \boldsymbol{q})\,] \left(S^2 + \frac{P^2}{m_\pi^2} \boldsymbol{q}^2 \right) .
\end{aligned}
$$ (7)

3.2 The Propagator Method. Mesonic Decay in the Medium

As a first approach to the decay in the medium, one could consider (4) but with nucleon and pion propagators properly modified by the presence of the medium. A simple way to do this is by considering nucleon occupation numbers given by a step function distribution, such that $n(\boldsymbol{k}) = 1$ for $\boldsymbol{k} \leq \boldsymbol{k}_F$ and 0 otherwise,

$$
\begin{aligned}
G(k - q) &\to \frac{1 - n(\boldsymbol{k} - \boldsymbol{q})}{(k^0 - q^0) - E(\boldsymbol{k} - \boldsymbol{q}) - V_N + i\epsilon} + \frac{n(\boldsymbol{k} - \boldsymbol{q})}{(k^0 - q^0) - E(\boldsymbol{k} - \boldsymbol{q}) - V_N - i\epsilon} \\
D(q) &\to \frac{1}{(q^0)^2 - \boldsymbol{q}^{\,2} - m_\pi^2 - \Pi(q^0, \boldsymbol{q})}
\end{aligned}
$$ (8)

with $\Pi(q^0, \boldsymbol{q})$ the pion self-energy. A common choice for the nucleon potential is $V_N = -k_F^2/(2M)$, which becomes r-dependent when the local Fermi momentum $k_F(r) = \left[\frac{3}{2}\pi^2\rho(r)\right]^{\frac{1}{3}}$ is used. Note that Coulomb effects have not been included in the evaluation of the in-medium Λ self-energy. One can easily compute the q^0–integration by performing a Wick rotation and choosing a contour integration as the one depicted in Fig. 5(a), where only the pole at the first quadrant, $q^0 = k^0 - E(\boldsymbol{k} - \boldsymbol{q}) - V_N$ contributes [18]. As pointed out in [18], the pole corresponding to $q^0 = k^0 - E(\boldsymbol{k} - \boldsymbol{q}) - V_N < 0$, which will contribute to the integral only when located in the third quadrant, is not considered because it corresponds to a very large value of $(\boldsymbol{k} - \boldsymbol{q})$, where the occupation number is zero, $n(\boldsymbol{k} - \boldsymbol{q}) = 0$.

The final result for the in-medium Λ width is:

$$
\begin{aligned}
\Gamma_\Lambda(k) &= -6(G_F m_\pi^2)^2 \int \frac{d^3q}{(2\pi)^3} [1 - n(\boldsymbol{k} - \boldsymbol{q})] \, \theta(k^0 - E(\boldsymbol{k} - \boldsymbol{q}) - V_N) \\
&\times \left(S^2 + \frac{P^2}{m_\pi^2} \boldsymbol{q}^2 \right) \text{Im} \left(\frac{1}{(q^0)^2 - \boldsymbol{q}^{\,2} - m_\pi^2 - \Pi(q^0, \boldsymbol{q})} \right)_{q^0 = k^0 - E(\boldsymbol{k} - \boldsymbol{q}) - V_N}
\end{aligned}
$$ (9)

As we have mentioned before, in finite nuclei is still possible to have mesonic decay since the Λ wave function has some overlap with the nuclear surface,

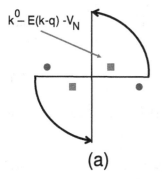
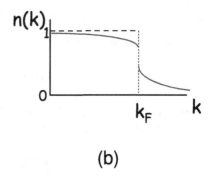

Fig. 5. (a) q^0-contour integration taken for the evaluation of the Λ width in the medium. The squares stand for the poles of the nucleon propagator, while the circles correspond to a renormalized pion energy. (b) Occupation number for an interacting Fermi sea of particles

where the Fermi momentum is smaller than $100\,\mathrm{MeV/c}$. Moreover, the momentum distribution of the Λ leads to spreading in the nucleon momenta allowing some of the nucleons to overcome the Pauli blocking.

Diagrammatically, the mesonic decay is given by the cuts shown in Fig. 4. The second diagram in the figure corresponds to the lowest order in the expansion of the pion propagator. Technically, the mesonic width can be obtained by subtracting the non-mesonic width from the total width (given above) or by replacing the pion pole in the evaluation of the free Λ width with the renormalized pion pole, obtained when we use the nucleon and pion propagators in the medium. In this latter case, the mesonic width is obtained from [15]:

$$\Gamma_\Lambda^{\mathrm{mes}} = 3(G_F m_\pi^2)^2 \int \frac{d^3q}{(2\pi)^3} \frac{1}{2\,\tilde{\omega}(q) - \dfrac{\partial \Pi}{\partial \tilde{\omega}}}$$

$$\times\, 2\pi\, \delta[\, E_\Lambda - \tilde{\omega}(q) - E_N(k-q) - V_N\,]\left(S^2 + \frac{P^2}{m_\pi^2}q^{\,2}\right),\quad (10)$$

where $\tilde{\omega}$ is the renormalized pion energy. From this expression, one can understand the possible increase on the mesonic width due to the attractive character of the pion self-energy, Π, which leads to a larger pion momentum for the same pion energy, and consequently, due to momentum conservation, to a larger momentum for the nucleon. All the previous expressions correspond to the nuclear matter formalism. To get the width for finite nuclei one typically uses a *Local Density Approximation* (LDA), according to which the Fermi momentum is local, through an explicit dependence on the local density,[3] typically $k_F(r) = \left[\frac{3}{2}\pi^2\rho(r)\right]^{\frac{1}{3}}$. The Fermi energy becomes then local

---

[3] A commonly used local density is given by the $\rho_A(r) = \dfrac{\rho_0}{\left(1+\exp(\frac{r-R(A)}{a})\right)}$, with $R(A) = 1.12\,A^{1/3} - 0.86\,A^{-1/3}$ and $a = 0.52\,\mathrm{fm}$.

through the relation $\varepsilon_F(\boldsymbol{r}) + V_N(\boldsymbol{r}) \equiv \dfrac{k_F^2(\boldsymbol{r})}{2\,m_N} + V_N(\boldsymbol{r}) = 0$, and the Λ width becomes momentum dependent,

$$\Gamma_\Lambda(\boldsymbol{k}) = \int d^3\boldsymbol{r}\, |\Psi_\Lambda(\boldsymbol{r})|^2\, \Gamma(\boldsymbol{k}, \rho(\boldsymbol{r})) \,. \tag{11}$$

Averaging over the momentum distribution of the Λ wave function gives finally the Λ width in the medium,

$$\Gamma_\Lambda = \int d^3\boldsymbol{k}\, |\tilde{\Psi}_\Lambda(\boldsymbol{k})|^2 \Gamma_\Lambda(\boldsymbol{k}) \,. \tag{12}$$

A refinement to the previous discussion is to consider, instead of simplistic step functions for the nucleon occupation number, distributions as in Fig. 5(b) more in agreement with an interacting Fermi sea, according to which, not all the states below the Fermi level are occupied, and not all the states above the Fermi level are empty. This refinement, instead of giving a more realistic estimation of the width, leads to an overestimation by 3 orders of magnitude for heavy nuclei [19]. The reason is that the above choice for the nucleon propagator is not the correct one. For an interacting Fermi sea of nucleons, the appropriate nucleon propagators have to be given in terms of spectral functions:

$$G(k^0, \boldsymbol{k}) = \int_{-\infty}^{\mu} d\omega \frac{S_h(\omega, k)}{k^0 - \omega - i\epsilon} + \int_{\mu}^{\infty} d\omega \frac{S_p(\omega, k)}{k^0 - \omega + i\epsilon} \,, \tag{13}$$

where μ is the chemical potential. Using this correct prescription one gets negligible modifications of the mesonic width for light-medium hypernuclei and significant improvement for heavy hypernuclei.

3.3 Wave Function Method. Mesonic Decay

The mesonic decay width in free space within this formalism is given by

$$\Gamma_\alpha^{\text{free}} = c_\alpha (G_F m_\pi^2)^2 \int \frac{d^3\boldsymbol{q}}{(2\pi)^3 2\,\omega(\boldsymbol{q})} 2\pi\, \delta[\, m_\Lambda - \omega(\boldsymbol{q}) - E_N \,]$$

$$\times \left(S^2 + \frac{P^2}{m_\pi^2}\, q^2 \right) \,, \tag{14}$$

where c_α is a constant which enforces the $\Delta I = 1/2$ rule, taking the value 1 for Γ_{π^0} or 2 for Γ_{π^-}. The constants S and P have the same meaning as in the previous sections. After performing the momentum integration one gets:

$$\Gamma_\alpha^{\text{free}} = c_\alpha (G_F m_\pi^2)^2 \frac{1}{2\pi} \frac{m_N\, q_{\text{c.m.}}}{m_\Lambda} \left(S^2 + \frac{P^2}{m_\pi^2} q_{\text{c.m.}}^2 \right) \,, \tag{15}$$

which accurately reproduces the experimental decay rate for the Λ hyperon in free space.

In a finite nucleus, (15) gets the contribution of some suppressing factors which represent the overlap of the pion, nucleon and Λ wave functions:

$$\Gamma_\alpha = c_\alpha (G_F m_\pi^2)^2 \sum_{N \notin F} \int \frac{d^3 q}{(2\pi)^3 2\,\omega(q)} \, 2\pi\, \delta[\, E_\Lambda - \omega(q) - E_N\,]$$

$$\times \left(S^2 \left| \int d^3 r \phi_\Lambda(r)\phi_\pi(q,r)\phi_N^*(r) \right|^2 + \frac{P^2}{m_\pi^2} \left| \int d^3 r \phi_\Lambda(r)\boldsymbol{\nabla}\phi_\pi(q,r)\phi_N^*(r) \right|^2 \right)$$

$$\times \left(S^2 + \frac{P^2}{m_\pi^2} q^2 \right) , \tag{16}$$

where the pion wave function is an outgoing wave, normalized to a plane wave at larger distances, which is a solution of the Klein-Gordon equation,

$$\left\{ \boldsymbol{\nabla}^2 - m_\pi^2 - 2\,\omega V_{opt}(r) + [\,\omega - V_C(r)\,]^2 \right\} \phi_\pi(q,r) = 0 \tag{17}$$

for a given energy eigenvalue $\omega = \omega(q)$. In the equation above, V_{opt} is the pion optical potential, related by $\Pi = 2\,\omega V_{opt}$ with the pion self-energy, and V_C introduces the Coulomb effects. It is through this equation that the mesonic decay shows a strong dependence on the pion-nucleus optical potential. The Λ and nucleon wave functions are obtainable through a shell model. Note that the sum in (16) is over non occupied nucleon orbitals.

To illustrate how well theoretical calculations compare with experimental numbers and how the PPM and WFM compare, we show in Table 1 the results for the mesonic decay width for $^5_\Lambda$He. The agreement one sees between

Table 1. Mesonic decay rate for $^5_\Lambda$He

Model	$\Gamma_M/\Gamma_\Lambda^{free}$	Ref.
PPM	0.65	Oset–Salcedo 1985 [18]
PPM	0.54	Oset–Salcedo–Usmani 1986 [24]
WFM	$0.331 \div 0.472$	Itonaga–Motoba–Bandō 1988 [22]
WFM (Quark Model wf)	0.608	Motoba et al 1991 [20]
WFM	0.61	Motoba 1992 [23]
WFM (Quark Model wf)	0.670	Straub et al 1993 [21]
WFM	0.60	Kumagai–Fuse et al 1995 [25]
	$0.59^{+0.44}_{-0.31}$	Exp BNL 1991 [26]
	0.541 ± 0.019	Exp KEK 2004 [27]

theoretical calculations and the BNL datum, does not show up when the calculations are compared with the more recent and accurate datum from KEK. These comparisons seem to indicate that a repulsive core in the Λ–α mean potential (used in all but the calculation of [18]) is favored. Note that this repulsion comes out automatically when the hypernuclear wave function is derived within a quark model [20, 21]. The results of [22, 23] refer to the use of different pion–nucleus optical potentials. Note that precise determinations of Γ_M are able to discriminate between different pion–nucleus optical potentials.

4 Finite Nucleus Calculation of the Non-Mesonic Decay

4.1 Polarization Propagator Method and Local Density Approximation

Following the same steps as in the mesonic mode, one extracts the Λ width from the imaginary part of the Λ self-energy, $\Gamma_\Lambda = -2\,\mathrm{Im}\Sigma_\Lambda$. Assuming an homogeneous system (nuclear matter) and within the Random Phase Approximation (RPA), which considers only one-particle/one-hole (1p–1h) excitations to describe residual two-body interactions between nucleons, not accounted for by mean-field potentials, the Λ self-energy reads [3]:

$$\Sigma_\Lambda(k) = 3\,\mathrm{i}\,(G_F m_\pi^2)^2 \int \frac{\mathrm{d}^4 q}{(2\pi)^4} \left(S^2 + \frac{P^2}{m_\pi^2}\, q^2 \right) F_\pi^2(q)\, G_N(k-q)\, G_\pi(q)\,, \quad (18)$$

where $F_\pi^2(q)$ is a (monopole) form factor which accounts for the hadronic structure at the $\pi\Lambda N$ vertex, and the nucleon and pion propagators read respectively:

$$G_N(p) = \frac{\theta(|\boldsymbol{p}| - k_F)}{p_0 - E_N(\boldsymbol{p}) - V_N + \mathrm{i}\varepsilon} + \frac{\theta(k_F - |\boldsymbol{p}|)}{p_0 - E_N(\boldsymbol{p}) - V_N - \mathrm{i}\varepsilon}\,,$$

$$G_\pi(q) = \frac{1}{q_0^2 - \boldsymbol{q}^2 - m_\pi^2 - \Pi_\pi(q)}\,. \quad (19)$$

4.2 Finite Nucleus Calculation

The direct evaluation of the non–mesonic decay rate in finite nuclei can be performed using the following expression:

$$\Gamma_1 = \int \frac{d^3 \boldsymbol{P}_{\mathrm{T}}}{(2\pi)^3} \int \frac{d^3 \boldsymbol{k}_{\mathrm{r}}}{(2\pi)^3} \, 2\pi \, \delta(M_{\mathrm{H}} - E_{\mathrm{R}} - E_1 - E_2)$$

$$\times \frac{1}{(2J+1)} \sum_{\substack{M_J \{R\} \\ \{1\}\{2\}}} |\mathcal{M}_{fi}|^2 \,, \tag{20}$$

where the subscript 1 stands for the one-nucleon induced channel. In the evaluation of the decay rate the initial hypernucleus, of mass M_{H}, is assumed to be at rest and one performs a sum over all quantum numbers of the final particles, i.e., the total energy of the residual nuclear system E_{R}, the total asymptotic energies of the emitted nucleons, $E_{1,2}$, as well as over the spin and isospin projections of the outgoing nucleons, $\{1\}$ and $\{2\}$, and residual system, $\{R\}$. The integration variables $\boldsymbol{P}_{\mathrm{T}} \equiv \boldsymbol{k}_1 + \boldsymbol{k}_2$ and $\boldsymbol{k}_{\mathrm{r}} \equiv (\boldsymbol{k}_1 - \boldsymbol{k}_2)/2$ are the total and relative momenta of the two outgoing nucleons. The momentum conserving delta function has been used to integrate out the momentum of the residual nucleus, $\boldsymbol{k}_{\mathrm{R}} = -\boldsymbol{P}_{\mathrm{T}}$. The factor $1/(2J+1)$, together with the sum, indicates an average over the initial hypernucleus total spin projections, M_J. Finally, $\mathcal{M}_{fi} = \langle \Psi_{\mathrm{R}}; \boldsymbol{P}_{\mathrm{T}} \boldsymbol{k}_{\mathrm{r}}, S \, M_S, T \, T_3 | \hat{O}_{\Lambda \mathrm{N} \to \mathrm{nN}} | \Psi_H \rangle$, is the amplitude for the transition from an initial hypernuclear state Ψ_H into a final state which is factorized into an anti-symmetrized two–nucleon state and a residual nuclear state Ψ_{R}. The two–nucleon state is characterized by the total momentum $\boldsymbol{P}_{\mathrm{T}}$, the relative momentum $\boldsymbol{k}_{\mathrm{r}}$, the spin and spin projection S, M_S and the isospin and isospin projection T, T_3. $\hat{O}_{\Lambda \mathrm{N} \to \mathrm{nN}}$ is a two–body transition operator acting on all possible $\Lambda \mathrm{N}$ pairs. In order to evaluate the two–body transition amplitude, one has to decouple from the initial hypernucleus a $\Lambda \mathrm{N}$ pair. To do this, a reasonable assumption is to use a weak coupling scheme of the Λ to the $(A-1)$-particles core, which consists in assuming that the Λ particle couples only to the core ground state,

$$|^{A}_{\Lambda} Z \rangle^{J_I M_I}_{T_I T_{3I}} = |\alpha_\Lambda \rangle \otimes |A-1\rangle$$

$$= \sum_{m_\Lambda M_C} \langle j_\Lambda m_\Lambda J_C M_C | J_I M_I \rangle \, |(n_\Lambda l_\Lambda s_\Lambda) j_\Lambda m_\Lambda \rangle \, |J_C M_C T_I T_{3I}\rangle. \tag{21}$$

Moreover, the technique of the coefficients of fractional parentage allows us to decouple from the $(A-1)$ antisymmetric core wave function one of the nucleons, leaving a properly anti-symmetrized residual $(A-2)$-particle system,

$$\Psi^{J_C T_C \alpha}_{\mathrm{as}}(1....N) = \sum_{J_{R_0} T_{R_0}} \sum_{\alpha_0 j_{\mathrm{N}}} \langle J_C T_C \alpha \{| J_{R_0} T_{R_0} \alpha_0, j_{\mathrm{N}} \rangle$$

$$\times [\Psi^{J_{R_0} T_{R_0} \alpha_0}_{\mathrm{as}}(1....N-1) \otimes \phi^{j_{\mathrm{N}}}(N)]_{J_C T_C}. \tag{22}$$

With all the previous techniques and working in a coupled two-body spin and isospin basis, the non-mesonic decay rate can be written as the sum of the neutron- ($\Lambda \mathrm{n} \to \mathrm{nn}$) and proton-induced ($\Lambda \mathrm{p} \to \mathrm{np}$) decay rates, $\Gamma_1 = \Gamma_{\mathrm{n}} + \Gamma_{\mathrm{p}}$. They are given by ($\mathrm{N} = \mathrm{n}, \mathrm{p}$):

$$\Gamma_N = \int \frac{d^3 \boldsymbol{P}_T}{(2\pi)^3} \int \frac{d^3 \boldsymbol{k}_r}{(2\pi)^3} 2\pi \, \delta(M_H - E_R - E_1 - E_2)$$

$$\times \sum_{S M_S} \sum_{J_R M_R} \sum_{T_R T_{3_R}} \frac{1}{2J+1} \sum_{M_J} |\langle T_R T_{3_R}, \tfrac{1}{2} t_{3N} \mid T_I T_{3I} \rangle|^2$$

$$\times \left| \sum_{T T_3} \langle T T_3 \mid \tfrac{1}{2} - \tfrac{1}{2}, \tfrac{1}{2} t_{3N} \rangle \sum_{m_\Lambda M_C} \langle j_\Lambda m_\Lambda, J_C M_C \mid J M_J \rangle \right.$$

$$\times \sum_{j_N} S^{1/2}(J_C T_I; J_R T_R, j_N t_{3N}) \times \sum_{M_R m_N} \langle J_R M_R, j_N m_N \mid J_C M_C \rangle \qquad (23)$$

$$\times \sum_{m_{l_N} m_{s_N}} \langle j_N m_N \mid l_N m_{l_N}, \tfrac{1}{2} m_{s_N} \rangle \times \sum_{m_{l_\Lambda} m_{s_\Lambda}} \langle j_\Lambda m_\Lambda \mid l_\Lambda m_{l_\Lambda}, \tfrac{1}{2} m_{s_\Lambda} \rangle$$

$$\times \sum_{S_0 M_{S_0}} \langle S_0 M_{S_0} \mid \tfrac{1}{2} m_{s_\Lambda}, \tfrac{1}{2} m_{s_N} \rangle \times \sum_{T_0 T_{30}} \langle T_0 T_{30} \mid \tfrac{1}{2} - \tfrac{1}{2}, \tfrac{1}{2} t_{3N} \rangle$$

$$\left. \times \frac{1 - (-1)^{(L+S+T)}}{\sqrt{2}} \, t_{\Lambda N \to nN}(S, M_S, T, T_3, S_0, M_{S_0}, T_0, T_{30}, l_\Lambda, l_N, \boldsymbol{P}_T, \boldsymbol{k}_r) \right|^2,$$

where $S^{1/2}(J_C T_I; J_R T_R, j_N t_{3N})$ is a nucleon pick–up spectroscopic amplitude, $t_{3p} = 1/2$ and $t_{3n} = -1/2$. The elementary amplitude $t_{\Lambda N \to nN}$ accounts for the transition from an initial ΛN state with spin (isospin) S_0 (T_0) to a final antisymmetric nN state with spin (isospin) S (T). It can be written in terms of other elementary amplitudes which depend on center-of-mass ("R") and relative ("r") orbital angular momentum quantum numbers of the ΛN and nN systems:

$$t_{\Lambda N \to nN} = \sum_{N_r L_r N_R L_R} X(N_r L_r, N_R L_R, l_\Lambda l_N) \, t_{\Lambda N \to nN}^{N_r L_r N_R L_R}, \qquad (24)$$

where the dependence on the spin and isospin quantum numbers has to be understood. In (24), the coefficients $X(N_r L_r, N_R L_R, l_\Lambda l_N)$ are the well known Moshinsky brackets, while:

$$t_{\Lambda N \to nN}^{N_r L_r N_R L_R} = \frac{1}{\sqrt{2}} \int d^3 \boldsymbol{R} \int d^3 \boldsymbol{r} \, e^{-i\boldsymbol{P}_T \cdot \boldsymbol{R}} \, \Psi_{\boldsymbol{k}_r}^*(\boldsymbol{r}) \, \chi_{M_S}^{\dagger S} \, \chi_{T_3}^{\dagger T}$$

$$\times V_{\sigma,\tau}(\boldsymbol{r}) \, \Phi_{N_R L_R}^{CM}\left(\frac{\boldsymbol{R}}{b/\sqrt{2}}\right) \Phi_{N_r L_r}^{rel}\left(\frac{\boldsymbol{r}}{\sqrt{2}b}\right) \chi_{M_{S_0}}^{S_0} \chi_{T_{30}}^{T_0}. \qquad (25)$$

Here, $V_{\sigma,\tau}(\boldsymbol{r})$ stands for the one-meson-exchange weak potential, which depends on the relative distance between the interacting Λ and nucleon as well as on their spin and isospin quantum numbers. Moreover, $\Phi_{N_r L_r}^{rel}(\boldsymbol{r}/(\sqrt{2}b))$ and $\Phi_{N_R L_R}^{CM}(\boldsymbol{R}/(b/\sqrt{2}))$ are the relative and center-of-mass harmonic oscillator wave functions describing the ΛN system, while $\Psi_{\boldsymbol{k}_r}(\boldsymbol{r})$ is the relative wave function of the nN final state.

The main point is now to decide which is the potential which will drive the weak two-body transition. In a meson-exchange model, this transition takes place through the virtual exchange of mesons belonging to the pseudoscalar and vector octets. Within this picture, the meson emitted at the weak vertex is viewed as absorbed by one of the nucleons in the medium, as depicted in Fig. 6. The leading pion contribution is given by the following weak and strong Lagrangians:

$$\mathcal{L}_{\Lambda N\pi}^{W} = -i\, G_F m_\pi^2\, \overline{\psi}_N (A_\pi + B_\pi \gamma_5)\, \boldsymbol{\tau} \cdot \boldsymbol{\phi}_\pi \psi_\Lambda \begin{pmatrix} 0 \\ 1 \end{pmatrix}$$
$$\mathcal{L}_{NN\pi}^{S} = -i\, g_{NN\pi}\, \overline{\psi}_N \gamma_5 \boldsymbol{\tau} \cdot \boldsymbol{\phi}_\pi \psi_N \,. \tag{26}$$

In (26), $g_{NN\pi} = 13.16$ is the coupling at the strong nucleon-nucleon-pion vertex, and the empirical constants, $A_\pi = 1.05$ and $B_\pi = -7.15$, have the same meaning as in the previous section. The nucleon, Λ and pion fields are given by ψ_N, ψ_Λ and $\boldsymbol{\phi}_\pi$, respectively, while the isospin spurion $\begin{pmatrix} 0 \\ 1 \end{pmatrix}$ is included to enforce the empirical $\Delta I = 1/2$ rule. The Bjorken and Drell convention for the definition of γ_5 [28] has been used. By inserting the pion propagator between the weak and strong vertices in (26) and performing the non relativistic reduction of the resulting Feynman amplitude, the momentum space transition potential for pion exchange (OPE) is given by:

$$V_{OPE}(\boldsymbol{q}) = -G_F m_\pi^2 \frac{g_{NN\pi}}{2M_S} \left(A_\pi + \frac{B_\pi}{2M_W} \boldsymbol{\sigma}_1 \boldsymbol{q} \right) \frac{\boldsymbol{\sigma}_2\, \boldsymbol{q}}{\boldsymbol{q}^2 + m_\pi^2} \boldsymbol{\tau}_1 \boldsymbol{\tau}_2\,, \tag{27}$$

where $\boldsymbol{q}$ represents the momentum transfer directed towards the strong vertex, m_π the pion mass and M_S (M_W) the average of the baryon masses at the strong (weak) vertex.

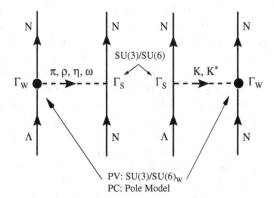

Fig. 6. The weak $\Lambda N \to NN$ transition proceeding through the virtual exchange of mesons belonging to the pseudoscalar and vector meson octets. By convention, the momentum is directed towards the strong vertex. The Parity-Violating weak baryon-baryon couplings are obtained by using SU(3) for pseudoscalar mesons and SU(6)$_W$ for vector mesons. The Parity-Conserving ones are obtained within the pole model formalism explained in the text

4.3 Neutron-to-Proton Ratio. Beyond
the One-Pion-Exchange Mechanism

Theoretically, and within the framework of the wave function method, one can evaluate separately the decay induced by a neutron Γ_n: $\Lambda n \to nn$ $(t_{3N} = -1/2)$ and the decay induced by a proton, Γ_p: $\Lambda p \to np$ $(t_{3N} = +1/2)$ and construct the ratio between both quantities, the neutron-to-proton ratio, Γ_n/Γ_p. Experimentally, one measures the final states, therefore, the nucleon(s) and any residual bound system left behind, hoping for a determination of the ratio between detected nn and np pairs, N_{nn}/N_{np}. For many years, theoretical evaluations of the former ratio gave very small numbers compared to the experimental extractions, which were quoted in the range: $0.5 \leq [\Gamma_n/\Gamma_p]^{Exp} \leq 2$. The smallness of the theoretical estimations is due to the tensor dominance in the weak pion-exchange mechanism. This tensor channel connects the $^3S_1$ ΛN state with the $^3D_1$ NN state, which is an isospin $I = 0$ state due to the antisymmetry of the NN wave function. $I = 0$ states can only happen in np pairs, suppressing the final two-neutron states and consequently, the neutron-to-proton ratio, giving values in the range $0.05 \leq [\Gamma_n/\Gamma_p]^{Theor} \leq 0.20$.

Different theoretical mechanisms were proposed to increase the value of the ratio, either through the decrease of Γ_p or the increase of Γ_n, with a destructive interference in the proton-induced channel or a constructive interference in the neutron-induced one. The large momentum transfer in the weak reaction indicates that short range effects could be important. This argument, in a meson exchange picture, is equivalent to include more massive mesons in the exchange mechanism to explore shorter distances. The ϱ meson (vector meson) is the isospin partner of the π (pseudoscalar meson), therefore, it seems reasonable to start by inserting this contribution first. Its inclusion was also motivated by the fact that, in the NN sector, the ϱ tensor transition interfered destructively with the tensor pion potential. In 1984 McKellar and Gibson included the ϱ meson in the weak reaction [29], in the framework of nuclear matter and considering only tensor transitions. The $\Lambda N \varrho$ coupling cannot be obtained from experiments, due to the lack of phase space to produce this meson on shell. To obtain this coupling, the authors used the factorization approximation, but this choice implied an ambiguity regarding the sign relative to the pion potential. Takeuchi, Takaki and Bandō [30] studied $A = 4$ and 5 systems with tensor transitions only, and used also factorization for the coupling of the ϱ to the baryons. Parreño et al [16] included all the possible transition channels besides the tensor one, and pointed out the relevance of a central spin-independent amplitude in the transition amplitude. The coupling constant was derived within SU(6)$_W$,[4] which allows us to write the unknown constant in terms of the pion coupling, the only one

---

[4] The SU(6)$_W$ group describes the product of the SU(3) flavor (u,d,s) group with the SU(2)$_W$ group associated with spin-1/2 fermions, with W the $W-$spin [17]. SU(2)$_W$ is preferred to SU(2), because boosts along the $z-$axis do not cause transitions between different representations of SU(2)$_W$, as it happens with SU(2).

accessible by experiments. The uncertainty regarding the relative sign between pion and rho disappeared. The final result is that the global interferences with the OPE mechanism produced a very negligible effect on the total rate and on the n/p ratio. A further sophistication of the OME model is the inclusion of the remaining pseudoscalar and vector mesons. The complete OME potential is obtained by considering the exchange of all the mesons with a mass up to 1 GeV, the π, ϱ, K, K*, η and ω mesons in the weak transition [9, 10, 11].

As we mentioned before, only the coupling constants of the baryons to a π meson are known experimentally. To get the couplings corresponding to the other mesons, a convenient procedure is to use flavor symmetry relations which allow us to write the amplitudes for the weak and strong vertices involving heavy mesons in terms of the experimentally known amplitudes involving the pion.

4.4 The Parity-Violating Amplitudes

The traditional approximation employed to obtain the PV amplitudes for the nonleptonic decays B $\rightarrow$ B$'$ + M has been the use of the soft-meson reduction theorem:

$$\lim_{q \to 0} \langle B'M_i(q)|H_{\mathrm{PV}}|B\rangle = -\frac{i}{F_\pi}\langle B'|[F_i^5, H_{\mathrm{PV}}]|B\rangle = -\frac{i}{F_\pi}\langle B'|[F_i, H_{\mathrm{PC}}]|B\rangle \,,$$

(28)

where q is the momentum of the meson and F_i is an SU(3) generator whose action on a baryon B_j gives $F_i|B_j\rangle = i f_{ijk}|B_k\rangle$. Since the weak Hamiltonian H_{W} is assumed to transform like the sixth component of an octet, a term like $\langle B_k|H_W^6|B_j\rangle$ can be expressed as:

$$\langle B_k|H_W^6|B_j\rangle = i\, F f_{6jk} + D\, d_{6jk} \,,$$

(29)

where f_{ijk} and d_{ijk} are the SU(3) coefficients and F and D the reduced matrix elements.

With the use of these soft-meson techniques and the SU(3) symmetry one can now relate the physical amplitudes of the nonleptonic hyperon decays into a pion plus a nucleon or a hyperon, B $\rightarrow$ B$'$ + π, with the unphysical amplitudes of the other pseudoscalar members of the meson octet, the kaon and the eta. One obtains relations such as [31]:

$$\langle nK^0|H_{\mathrm{PV}}|n\rangle = \sqrt{\frac{3}{2}}\,\Lambda_-^0 - \frac{1}{\sqrt{2}}\,\Sigma_0^+ \,, \quad \langle pK^0|H_{\mathrm{PV}}|p\rangle = -\sqrt{2}\,\Sigma_0^+ \,, \quad (30)$$

$$\langle nK^+|H_{\mathrm{PV}}|p\rangle = \sqrt{\frac{3}{2}}\,\Lambda_-^0 + \frac{1}{\sqrt{2}}\,\Sigma_0^+ \,, \quad \langle n\eta|H_{\mathrm{PV}}|\Lambda\rangle = \sqrt{\frac{3}{2}}\,\Lambda_-^0 \,, \quad (31)$$

where Σ_0^+ (Λ_-^0) stands for $\langle p\pi^0|H_{\mathrm{PV}}|\Sigma^+\rangle$ ($\langle p\pi^-|H_{\mathrm{PV}}|\Lambda\rangle$), the PV amplitude of the decay $\Sigma^+ \rightarrow p\pi^0$ ($\Lambda \rightarrow p\pi^-$), which is experimentally accessible. In all these expressions the standard notation has been used, according to which the hyperon and meson charges appear as superscript and subscript, respectively. An explicit calculation of the parity-violating $\langle nK^+|H_{PV}|p\rangle$ amplitude is shown in the Appendix of [32].

While SU(3) symmetry allows connecting the amplitudes of the physical pionic decays with those of the unphysical decays involving etas and kaons, $SU(6)_W$ permits to relate the amplitudes involving pseudoscalar mesons with those of the vector mesons. The calculation of these couplings is tedious and is out of the scope of the present lectures. References [9, 10, 31, 32] give more details as well as the final results for these couplings.

4.5 The Parity-Conserving Amplitudes

A description of the physical nonleptonic decay amplitudes $B \to B' + \pi$ can also be performed by using a lowest-order chiral analysis. Employing a chiral Lagrangian truncated at lowest order in the energy expansion for the PV (or s-wave) amplitudes, yields results identical to those discussed above for pseudoscalar mesons. However, if one defines the lowest-order chiral Lagrangian for PC (or p-wave) amplitudes, one finds that such an operator has to vanish since it has the wrong transformation properties under CP symmetry. Thus, the only allowed chiral Lagrangian at lowest order can generate PV but not PC terms. The standard method to compute the PC amplitudes is the so-called pole model. As shown in [33], this approach can be motivated by considering the transition amplitude for the nonleptonic emission of a meson

$$\langle B'M_i(q)|H_W|B\rangle = \int d^4x\, e^{iqx}\, \theta(x^0)\, \langle B'|[\partial A_i(x), H_W(0)]|B\rangle\,, \qquad (32)$$

where A_i is the weak current relevant for the transition. Inserting a complete set of intermediate states, $\{|n\rangle\}$, one can show that

$$\langle B'M_i(q)|H_W|B\rangle = -\int d^3x\, e^{iqx}\, \langle B'|[A_i^0(x,0), H_W(0)]|B\rangle - q_\mu M_i^\mu\,, \qquad (33)$$

where

$$\begin{aligned}M_i^\mu &= (2\pi)^3 \sum_n \Bigg[\delta(\boldsymbol{p}_n - \boldsymbol{p}_{B'} - \boldsymbol{q})\frac{\langle B'|A_i^\mu(0)|n\rangle\langle n|H_W(0)|B\rangle}{p_B^0 - p_n^0}\\&+ \delta(\boldsymbol{p}_B - \boldsymbol{p}_n - \boldsymbol{q})\frac{\langle B'|H_W(0)|n\rangle\langle n|A_i^\mu(0)|B\rangle}{p_B^0 - q^0 - p_n^0}\Bigg]\,.\end{aligned} \qquad (34)$$

While the first term in (33) becomes the commutator introduced in (28), the second term contains contributions from the $\frac{1}{2}^+$ ground state baryons which are singular in the SU(3) soft-meson limit. These pole terms become the leading contribution to the PC amplitudes. Note in passing that in principle, such baryon-pole terms can also contribute to the PV amplitudes, however, more detailed studies [33] showed that their magnitude is only several per cent of the leading current algebra contribution. The baryon-pole diagrams contributing to the PC vertices for the exchange of the π and ϱ mesons are depicted in Fig. 7. To illustrate the method, I explicitly give the expression for the p-wave amplitude of the $\Lambda \to N\pi$ decay, where one can actually compare with experiment. The contribution to the PC weak vertex coming from the baryon-pole diagrams are given by:

$$B_\pi = g_{\mathrm{NN}\pi} \frac{1}{m_\Lambda - m_\mathrm{N}} A_{\mathrm{N}\Lambda} + g_{\Lambda\Sigma\pi} \frac{1}{m_\mathrm{N} - m_\Sigma} A_{\mathrm{N}\Sigma} , \tag{35}$$

where $A_{\mathrm{N}\Lambda}$ and $A_{\mathrm{N}\Sigma}$ are weak baryon $\to$ baryon transition amplitudes that can be related to the processes $\Lambda \to N\pi$ and $\Sigma \to N\pi$. These quantities can be determined via current algebra/PCAC as before

$$\lim_{q \to 0} \langle \pi^0 n | H_{\mathrm{PV}} | \Lambda \rangle = \frac{-i}{F_\pi} \langle n | [F_{\pi^0}^5, H_{\mathrm{PV}}] | \Lambda \rangle = \frac{i}{2F_\pi} \langle n | H_{\mathrm{PC}} | \Lambda \rangle \tag{36}$$

$$\lim_{q \to 0} \langle \pi^0 p | H_{\mathrm{PV}} | \Sigma^+ \rangle = \frac{-i}{F_\pi} \langle p | [F_{\pi^0}^5, H_{\mathrm{PV}}] | \Sigma^+ \rangle = \frac{i}{2F_\pi} \langle p | H_{\mathrm{PC}} | \Sigma^+ \rangle . \tag{37}$$

Then assuming no momentum dependence for the baryon s-wave decay amplitude and absorbing the i factor in the definitions of $A_{\mathrm{N}\Lambda}$ and $A_{\mathrm{N}\Sigma}$, one gets:

$$A_{\mathrm{N}\Lambda} = i \langle n | H_{\mathrm{PC}} | \Lambda \rangle = 2F_\pi \langle \pi^0 n | H_{\mathrm{PV}} | \Lambda \rangle = -\sqrt{2} F_\pi \langle \pi^- p | H_{\mathrm{PV}} | \Lambda \rangle$$
$$= -4.32 \times 10^{-5} \text{ MeV} \tag{38}$$

$$A_{\mathrm{N}\Sigma} = \frac{i}{\sqrt{2}} \langle p | H_{\mathrm{PC}} | \Sigma^+ \rangle = \sqrt{2} F_\pi \langle \pi^0 p | H_{\mathrm{PV}} | \Sigma^+ \rangle$$
$$= -4.35 \times 10^{-5} \text{ MeV} . \tag{39}$$

With these values, obtained from the physical $\Lambda \to p\pi^-$ and $\Sigma^+ \to p\pi^0$ parity-violating amplitudes, and using the Nijmegen strong coupling constants $g_{\mathrm{NN}\pi}$ and $g_{\Lambda\Sigma\pi}$, one derives $B_\pi = -11.98 \times 10^{-7}$, which is within 24% of the

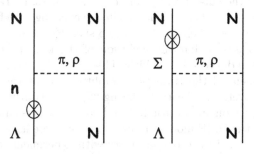

Fig. 7. Baryon-pole diagrams contributing to the PC weak vertices in the $\Lambda N \to NN$ transition amplitude for the exchange of the isovector π and ϱ mesons

experimental value ($B_\pi^{\text{exp}} = -7.15 \times G_F m_\pi^2 = -15.80 \times 10^{-7}$). If one chooses the Jülich B strong couplings rather than the Nijmegen ones, the new value for B_π is -15.74×10^{-7}, closer to the experimental one. In all calculations though, the experimental value is taken.

4.6 The Weak $\Lambda N \to NN$ Transition Potential

The final expression for the one-meson-exchange potential including the six mesons is:

$$V(\boldsymbol{r}) = \sum_i \sum_\alpha V_\alpha^{(i)}(\boldsymbol{r}) = \sum_i \sum_\alpha V_\alpha^{(i)}(r) \hat{O}_\alpha(\hat{r}, \boldsymbol{\sigma}) \hat{I}_\alpha^{(i)} \qquad (40)$$

where the index i ($= 1, ...6$) stands for the different mesons exchanged and the index α ($\alpha = 1, ...4$) for the different transition channels (central spin-independent, central spin-dependent, tensor and parity violating). The spin-angular dependence of the potential is contained in the $\hat{O}_\alpha(\hat{r}, \boldsymbol{\sigma})$ operator, which reads: $\hat{1}$, $\boldsymbol{\sigma}_1 \boldsymbol{\sigma}_2$, $S_{12}(\hat{r}) = 3\,\boldsymbol{\sigma}_1 \hat{r} \boldsymbol{\sigma}_2 \hat{r} - \boldsymbol{\sigma}_1 \boldsymbol{\sigma}_2$, and $\boldsymbol{\sigma}_2 \hat{r}$ (for pseudoscalar mesons) or $[\boldsymbol{\sigma}_1 \times \boldsymbol{\sigma}_2] \hat{r}$ (for vector mesons) for $\alpha = 1, 2, 3$, and 4 respectively. The operator $\hat{I}_\alpha^{(i)}$ contains, apart from the particular isospin structure corresponding to the meson exchanged ($\hat{1}$, $\boldsymbol{\tau}_1 \boldsymbol{\tau}_2$ or a combination of both), the weak baryon-baryon-meson coupling for each PC and PV amplitude. To regularize the potential at short distances, i.e. to account for finite size effects, the usual procedure in meson-exchange models is to include form factors:

$$V(\boldsymbol{r}) = \int \frac{\mathrm{d}^3 \boldsymbol{q}}{(2\pi)^3} \frac{e^{i \boldsymbol{q} \boldsymbol{r}}}{\boldsymbol{q}^2 + \mu^2 - q_0^2} \tilde{V}(\boldsymbol{q}) F^2(\boldsymbol{q}^2). \qquad (41)$$

Typical choices for $F^2(\boldsymbol{q}^2)$ are the dipole, monopole and Gaussian form factors:

$$F^2(\boldsymbol{q}^2) = \left(\frac{\tilde{\Lambda}^2 - \mu^2}{\tilde{\Lambda}^2 + \boldsymbol{q}^2} \right)^2, \; F^2(\boldsymbol{q}^2) = \left(\frac{\tilde{\Lambda}^2 - \mu^2}{\tilde{\Lambda}^2 + \boldsymbol{q}^2} \right), \; F^2(\boldsymbol{q}^2) = \exp \left(-\frac{\boldsymbol{q}^2}{\tilde{\Lambda}^2} \right), \qquad (42)$$

where $\tilde{\Lambda}$ stands for the cut-off and μ for the meson mass. In Table 2 we illustrate the effects of adding heavier mesons in the exchange mechanism for two typical hypernuclei, the s-shell $^5_\Lambda$He and the p-shell $^{12}_\Lambda$C. From these numbers, taken from [11], one sees that the inclusion of the kaon produces a reduction of the proton induced decay width. This is due to the opposite sign in the tensor component of the transition amplitude of the kaon with respect to the pion. Moreover, the Parity-Violating $^3S_1 \to {}^3P_1$ transition, which contributes to both, neutron and proton induced processes, gets enhanced by kaon exchange. Besides the OME model discussed above, there have been other approaches that have tried to reconcile theory with experiments during the last years. One of the first attempts was carried out by M. Shmatikov [34] and by K. Itonaga et al [35], by considering the contribution of two correlated pions in

Table 2. Non-mesonic decay rate in units of $\Gamma_\Lambda^{\text{free}}$ and neutron-to-proton ratio predictions of the one-meson-exchange model for $^5_\Lambda$He and $^{12}_\Lambda$C. The first row shows the results for the pion-exchange mechanism, the second row shows the effect of including the kaon-exchange on top of the OPE mechanism, while the last row shows the final result obtained by the inclusion of all six mesons

	$^5_\Lambda$He		$^{12}_\Lambda$C	
	Γ_{nm}	$\Gamma_{\text{n}}/\Gamma_{\text{p}}$	Γ_{nm}	$\Gamma_{\text{n}}/\Gamma_{\text{p}}$
π	0.43	0.09	0.75	0.08
$\pi + \text{K}$	0.24	0.50	0.41	0.34
all mesons	0.32	0.46	0.55	0.34

the exchange mechanism. These pions can couple to an isoscalar "σ"-like state or to an isovector "ϱ"-type state. While the scalar-isoscalar channels showed a strong central transition component, the vector-isovector one showed a strong tensor component. M. Shmatikov found cancellation between the diagrams involving a Σ or a nucleon as intermediate state, resulting in a significant $J = 0$ contribution. Jido et al [36] used the language of propagators to compute the pion, kaon and correlated two-pion contributions to the Λ self-energy in the medium. Still, another approach considered the combination of the long ranged π and K with an Effective Quark Hamiltonian which automatically incorporates $\Delta I = 3/2$ transitions [12, 37]. As it is well known, QCD corrections to the basic weak interactions produce an effective weak Hamiltonian, $\mathcal{H}_{\text{eff}}$, which can be evaluated using perturbative QCD, down to a scale $\sim 1\,\text{GeV}$. The form of $\mathcal{H}_{\text{eff}}$ for the non-leptonic strangeness changing weak interactions is then found to be [38, 39, 40]:

$$\mathcal{H}_{\text{eff}} = -\sqrt{2}\, G_{\text{F}} \sin \theta_{\text{C}} \cos \theta_{\text{C}} \sum_{i=1}^{6} c_i O_i \,, \tag{43}$$

where θ_{C} the Cabbibo angle and the operators O_i have the form:

$$
\begin{aligned}
O_1 &= \bar{d}_L \gamma_\mu s_L\, \bar{u}_L \gamma^\mu u_L - \bar{u}_L \gamma_\mu s_L\, \bar{d}_L \gamma^\mu u_L \\
O_2 &= \bar{d}_L \gamma_\mu s_L\, \bar{u}_L \gamma^\mu u_L + \bar{u}_L \gamma_\mu s_L\, \bar{d}_L \gamma^\mu u_L + 2\bar{d}_L \gamma_\mu s_L\, \bar{d}_L \gamma^\mu d_L \\
&\quad + 2\bar{d}_L \gamma_\mu s_L\, \bar{s}_L \gamma^\mu s_L \\
O_3 &= \bar{d}_L \gamma_\mu s_L\, \bar{u}_L \gamma^\mu u_L + \bar{u}_L \gamma_\mu s_L\, \bar{d}_L \gamma^\mu u_L + 2\bar{d}_L \gamma_\mu s_L\, \bar{d}_L \gamma^\mu d_L \\
&\quad - 3\bar{d}_L \gamma_\mu s_L\, \bar{s}_L \gamma^\mu s_L \\
O_4 &= \bar{d}_L \gamma_\mu s_L\, \bar{u}_L \gamma^\mu u_L + \bar{u}_L \gamma_\mu s_L\, \bar{d}_L \gamma^\mu u_L - \bar{d}_L \gamma_\mu s_L\, \bar{d}_L \gamma^\mu d_L \\
O_5 &= \bar{d}_L \gamma_\mu \lambda^a s_L\, (\bar{u}_R \gamma^\mu \lambda^a u_R + \bar{d}_R \gamma^\mu \lambda^a d_R + \bar{s}_R \gamma^\mu \lambda^a s_R) \\
O_6 &= \bar{d}_L \gamma_\mu s_L\, (\bar{u}_R \gamma^\mu u_R + \bar{d}_R \gamma^\mu d_R + \bar{s}_R \gamma^\mu s_R) \,.
\end{aligned}
\tag{44}
$$

Table 3. Non–mesonic decay rate for $^{5}_{\Lambda}$He and $^{12}_{\Lambda}$C in units of the free Λ decay width. Different theoretical estimates are compared to the available experimental data

$^{5}_{\Lambda}$He	$^{12}_{\Lambda}$C	Model and Reference
0.5		WFM: OPE + 4BPI [43]
	1.28	WFM: hybrid [44]
1.15	1.5	PPM: Correlated OPE [18]
0.54		PPM: Correlated OPE [24]
0.519		WFM: π + K+ DQ [45]
0.426	1.174	WFM: OPE + 4BPI [41]
	0.769	PPM: π + K + 2π + ω [36]
$0.317 \div 0.425$	$0.554 \div 0.726$	WFM: $\pi + \varrho + K + K^{*} + \omega + \eta$ [11]
0.422	1.060	WFM: $\pi + 2\pi/\varrho + 2\pi/\sigma + \omega$ [35]
0.44	0.93	OPE + OKE + 4BPI [42]
0.41 ± 0.14	1.14 ± 0.20	Exp BNL 1991 [26]
	0.89 ± 0.18	Exp KEK 1995 [46]
0.50 ± 0.07		Exp KEK 1995 [47]
	0.83 ± 0.11	Exp KEK 2000 [48, 49]
0.406 ± 0.020	0.953 ± 0.032	Exp KEK 2004 [27]

Table 4. Γ_{n}/Γ_{p} ratio for $^{5}_{\Lambda}$He and $^{12}_{\Lambda}$C. Different theoretical estimates are compared to the available experimental data

$^{5}_{\Lambda}$He	$^{12}_{\Lambda}$C	Model and Reference
0.701		π + K+ DQ [45]
	0.53	π + K + 2π + ω [36]
$0.343 \div 0.457$	$0.288 \div 0.341$	$\pi + \varrho + K + K^{*} + \omega + \eta$ [11]
0.386	0.368	$\pi + 2\pi/\varrho + 2\pi/\sigma + \omega$ [35]
0.55	0.77	OPE + OKE + 4BPI [42]
0.93 ± 0.55	$1.33^{+1.12}_{-0.81}$	BNL 1991 [26]
	$1.87^{+0.67}_{-1.16}$	KEK 1995 [46]
1.97 ± 0.67		KEK 1995 [47]
$0.45 \pm 0.11 \pm 0.03$	$0.51 \pm 0.13 \pm 0.05$	KEK 2004 [50, 51]

The Wilson coefficients, c_i, are scale-dependent and calculable perturbatively. The operators $O_1, \cdots, O_6$ have the specific $(flavor, isospin)$ quantum numbers $(8, 1/2)$, $(8, 1/2)$, $(27, 1/2)$, $(27, 3/2)$, $(8, 1/2)$ and $(8, 1/2)$, respectively. The operators $O_{5,6}$, with LR chiral structure are generated by QCD penguin-type corrections and, as noted above, have different chiral structure than do the remaining operators. Of the operators, $O_{1,\cdots,6}$, only O_4 is $\Delta I = 3/2$. This model uses the experimental baron-baryon-pion vertices while the vertices corresponding to coupling a kaon are obtained through SU(3) values. Finite size effects are also included through a monopole form-factor at each vertex to regularize the transition potential.

Tables 3 and 4 show the theoretical results of the calculation of the total one-nucleon induced decay rate and the Γ_n/Γ_p ratio for $^5_\Lambda$He and $^{12}_\Lambda$C, as compared to the available experimental data. The numerical values have been taken from the compilation of [3]. Note that the results corresponding to [41, 42] cannot be considered as pure theoretical models, since those are effective approaches which include a series of short-range contact terms whose coefficients are adjusted to reproduce some decay observables.

5 Baryon-Baryon Wave Functions

In this section we sketch how the strong interaction between the hadrons in the initial and final state can be accounted for in our formalism. This is vital when extracting information on the elementary weak two-body interaction taking place in the medium. Ideally, one should solve exactly the A-body wave function, which for most of the studied systems is not feasible. For bound systems including a hyperon, only the (finite nucleus) $^3_\Lambda$H wave function has been derived exactly [52], with the input of realistic potential models, in particular with the NSC97 YN [53] model. Cluster-type calculations have been also applied to obtain the wave function of light bound strange systems [54]. In light of this situation, an accurate method to describe strange bound systems is to perform a microscopic finite nucleus G-matrix calculation with the input of realistic potential models. The G-matrix formulation accounts for the strong interaction between two particles in the medium, and allows their propagation from occupied to unoccupied states in a shell-model picture. This method can be very involved for A > 5 and, in practice, one looks for simpler ways to account for the strong interaction in the initial system.

A convenient choice to determine the single-particle Λ and N orbits is a harmonic oscillator mean field potential, where the oscillator parameters b_Λ and b_N are adjusted to reproduce the experimental binding energies of the hypernucleus under consideration and the charge form factor of the residual core. For $^{12}_\Lambda$C and $^{11}_\Lambda$B one obtains $b_N = 1.64$ fm and $b_\Lambda = 1.87$ fm, while for $^5_\Lambda$He $b_N = 1.4$ fm and $b_\Lambda = 1.85$ fm. The mean-field two-particle wave functions are modified by short-range correlations generated by short-range nuclear forces. The correlated ΛN wave function is obtained from a G-matrix calculation [55] for $^5_\Lambda$He performed with the input of the soft-core and hard-core Nijmegen models [56]. The wave functions obtained in this way, for the singlet and triplet channels, are divided by our mean-field harmonic oscillator wave functions. With this strategy we produce effective correlation functions, $f_{\Lambda N}$, for the $^1S_0$ and $^3S_1$ ΛN states and for both, a hard- and a soft-core potential model. Our new ΛN wave function is finally obtained by multiplying the mean-field wave function by a spin independent parametrization which lies in between the hard- and soft-core results. The same correlation function is applied to s- and p-shell hypernuclei, and has the form:

$$f(r) = \left(1 - e^{-(r/a)^2}\right)^n + b\,r^2 e^{-(r/c)^2},$$

with $a = 0.5\,\text{fm}$, $b = 0.25\,\text{fm}$, $c = 1.28\,\text{fm}$ and $n = 2$. To illustrate this simplification and for the sake of comparison, this parametrization is depicted in Fig. 8 together with the hard-core and soft-core correlation functions obtained from the G-matrix calculation of [55] and for both spin channels.

In the absence of strong correlations the NN wave functions are just plane waves. Since the momentum transfer in the weak reaction is large, and therefore, the two outgoing nucleons are fast, one can take the approximation of considering only the mutual influence between both nucleons, and disregard the interaction of these nucleons with the rest of the system. The wave function describing the relative motion of two-particles moving under the influence of a two-body potential V, is obtained from the Lippmann-Schwinger equation:

$$|\,\Psi^{(\pm)}\rangle = |\,\phi\,\rangle + \frac{1}{E - H_0 \pm i\varepsilon}\,V\,|\,\Psi^{(\pm)}\rangle. \tag{45}$$

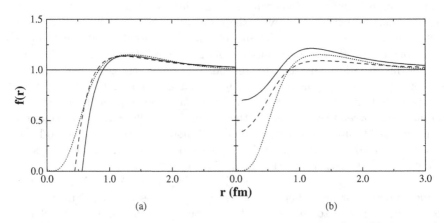

Fig. 8. The spin-independent parametrization discussed in the text (*dotted line*) compared to the correlation function obtained from the G-matrix calculation of [55] with the use of the hard-core (*left panel*) and soft-core (*right panel*) Nijmegen potentials of [56]. The *solid and dashed line* stand for the singlet and triplet ΛN channels respectively

Alternatively, one can write:

$$| \Psi^{(\pm)} \rangle = | \phi \rangle + \frac{1}{E - H_0 \pm i\varepsilon} T | \phi \rangle , \qquad (46)$$

where the T matrix, defined by: $T | \phi \rangle = V | \Psi^{(+)} \rangle$, $\langle \Psi^{(-)} | V = \langle \phi | T$, obeys:

$$T = V + V \frac{1}{E - H_0 + i\varepsilon} T . \qquad (47)$$

Projecting (45) into coordinate space, performing a partial-wave decomposition of the wave function and of the T-matrix elements, working in the coupled $(LS)J$ basis, and doing a non negligible amount of algebra, one finally obtains an expression for the correlated NN wave function:

$$\Psi_{L'S',LS}^{(-)* J}(k,r) = j_L(kr) \delta_{LL'} \delta_{SS'}$$
$$+ \int k'^2 dk' \frac{\langle k(LS)JM \mid T \mid k'(L'S')JM \rangle j_{L'}(k'r)}{E(k) - E(k') + i\eta} , \qquad (48)$$

where the partial wave T-matrix elements fulfill the integral equation:

$$\langle k(LS)JM \mid T \mid k'(L'S')JM \rangle = \langle k(LS)JM \mid V \mid k'(L'S')JM \rangle$$
$$+ \sum_{S''L''} \int k''^2 dk'' \frac{\langle k(LS)JM \mid V \mid k''(L''S'')JM \rangle}{E(k) - E(k'') + i\eta}$$
$$\times \langle k''(L''S'')JM \mid T \mid k'(L'S')JM \rangle . \qquad (49)$$

As input of the T-matrix equation one uses realistic potential models, as the ones provided by the Nijmegen group [53] or the Bonn-Jülich groups [57, 58]. To illustrate the effects of the different NN wave functions on the total and partial decay rates, we present the results of Table 5. The first row shows the values obtained in the absence of short-range correlations, the second row shows the results obtained when a Bessel type correlation function is used, and the third row shows the results corresponding to solving a

Table 5. Comparison of the theoretical predictions for the non-mesonic decay rate and the neutron-to-proton ratio obtained when one uses different prescriptions for the NN wave functions, i.e. to account for the strong NN interaction. The numbers are from [11]

$^5_\Lambda$He	$\Gamma_{nm}/\Gamma_\Lambda$	Γ_n/Γ_p
plane waves	0.72	0.61
$f(r) = 1 - j_0(q_c r), q_c = 3.93\,\mathrm{fm}$	0.77	0.62
T−matrix with NSC97f	0.32	0.46

T-matrix equation with the input of the Nijmegen Soft-Core model version 97f [53] (NSC97f). One can see that the omission of correlations for the final NN system, and even the consideration of phenomenological correlations instead of a more accurate T-matrix formalism, produces rates and Γ_n/Γ_p ratios fictitiously large.

6 Final State Interaction Effects

Table 4 showed that pure theoretical models are far from reproducing the experimental values for the neutron-to-proton ratio. Neither the inclusion of short range physics in the modeling of the weak $|\Delta S| = 1$ mechanism, nor the consideration of more realistic NN wave functions for the two final nucleons, can bring theory and experiment close enough. In this section we will see how this can be achieved by accounting for the propagation of the two emitted nucleons through the medium, and we will see how this fact influences the extraction of the ratio from the experimentally measured quantities. We know that experiments cannot measure directly this ratio, but some relation between detected neutrons and protons, $[N_n/N_p]^{exp}$, or detected nn pairs and np pairs, $[N_{nn}/N_{np}]^{exp}$. Some works have been devoted to establish how this experimental quantity, affected by the strong interaction among the nucleons, is related to the Γ_n/Γ_p ratio [59]. Primary nucleons may re scatter and undergo charge-exchange reactions with the rest of the nucleons in the medium. This effect can result in changes in their momentum, direction and charge. It is clear then that the correct comparison of theoretical results to decay observables has to include the effects of Final State Interactions (FSI). Moreover, the correct analysis has to include the 2N-induced channel too, since it can definitely affect the total number of neutrons and protons in the final state [60]. Since each n-induced process produces two neutrons and each p-induced process produces one neutron and one proton, if we only had one-nucleon induced processes, the number of protons per non-mesonic weak decay would be exactly $N_p = \Gamma_p$, while the total number of neutrons would be $N_n = 2\Gamma_n + \Gamma_p$. The 2N-induced mechanism, dominated by the np-induced reaction, clearly modifies both quantities. All these considerations have been reviewed in the framework of a finite nucleus calculation in [13, 14], where an intranuclear cascade calculation produced energy spectra for the detection of NN pairs in coincidence, as well as spectra for the angular correlations, in nice agreement with the most recent experimental data from KEK, when the OME prediction of [10, 11] for the hypernuclei was used. Those references use a modified version of the classical Monte Carlo of [59] for the study of hypernuclear decay in infinite nuclear matter, where the finite nucleus result was obtained through a Local Density Approximation. The approach includes also the 2N-induced mechanism as dominated by the absorption on a np-correlated pair at the weak $\Lambda n\pi$ vertex. Its contribution was estimated within the polarization propagator method in LDA [5, 61] to be around 20% for s-shell nuclei and 25% for p-shell ones. In [62], the authors used a microscopic model for the 2N-induced channel and evaluated the contribution of all three decay channels: nn, np and pp. They confirmed the dominance of the np channel and quoted similar percentages with respect to the total decay rate. The intranuclear cascade calculation will follow the fate of the 3 nucleons emitted afterwards.

The basic ingredients of the intranuclear cascade code are: i) a random generator produces primary nucleons at some particular point in space, position inside the nucleus, and with a particular value of momentum and charge, ii) the same random generator determines the decay channel, n-, p- or np-induced, according to their respective probabilities given by the values predicted by the full OME model in finite nucleus of [11], Γ_n, Γ_p, and the properly scaled value of Γ_{np}.[5] After the primary nucleons are emitted, they move under a local potential, $V_N(R) = -\dfrac{k_{F_N}^2(R)}{2\,m_N}$, with $k_{F_N}(R)$ the local Fermi momentum. The nucleons then collide with other nucleons of the medium according to NN cross sections corrected by Pauli blocking, producing among other effects, the emission of secondary nucleons. Therefore, each Monte Carlo event will end up with a certain number of nucleons which will leave the nucleus with some defined momentum and energies.

The calculated spectra for the number of protons emitted per non-mesonic weak decay for $^5_\Lambda$He and $^{12}_\Lambda$C are depicted in Fig. 9. Note that one has to be cautious when talking about Monte Carlo techniques applied to light systems like helium, and therefore, take the results for $^5_\Lambda$He as less realistic than the ones presented for $^{12}_\Lambda$C. The dashed line, which corresponds to the energy distribution of primary protons, includes the effects of the local potential but it does not consider their collisions with the nucleons in the medium. The calculations have been performed with the OME-f model, which takes the strong interaction ingredients (strong coupling constants, NN potential, etc.) from the Nijmegen soft-core model version f [53]. This model predicts $\Gamma_n/\Gamma_p = 0.46$ for $^5_\Lambda$He and $\Gamma_n/\Gamma_p = 0.34$ for $^{12}_\Lambda$C. The distributions present a peak around the most probable kinematics, which corresponds to the situation in which the two outgoing (fast) nucleons leave the system in opposite directions (back-to-back kinematics). This peak smears out when one takes into account FSI, since their effect is to produce secondary nucleons, as a result of one or multiple collisions, which therefore leave the system with smaller momenta. These secondary nucleons populate the low-energy region and as expected, FSI have a larger effect on the heavier system, $^{12}_\Lambda$C, than in the lighter one, $^5_\Lambda$He.

---

[5] The calculation mixes two different formalisms, the 1N-induced OME formalism in finite nuclei of [11], and the Polarization Propagator Method in LDA to determine the 2N-induced channel. This implies that the obtained distributions of the weak decay nucleons and the Γ_{2N} value have been properly normalized to keep the $\dfrac{\Gamma_{2N}}{\Gamma_{1N}}$ unchanged, $\dfrac{\Gamma_{2N}}{\Gamma_{1N}} \equiv \left(\dfrac{\Gamma_{2N}}{\Gamma_{1N}}\right)^{LDA} = 0.20$ for $^5_\Lambda$He and 0.25 for $^{12}_\Lambda$C.

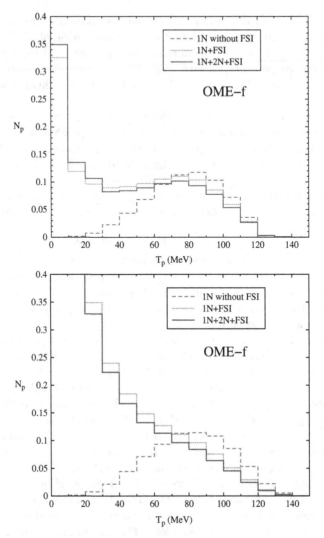

Fig. 9. Single proton distributions when accounting for Final State Interactions and for the 2N-induced channel as compared with the distribution of primary protons, for $^{5}_{\Lambda}$He (*upper panel*) and $^{12}_{\Lambda}$C (*lower panel*)

The single-neutron spectrum for $^{12}_{\Lambda}$C observed in the KEK–E369 experiment [63] is well reproduced by the theoretical calculations, as can be seen from Fig. 10, where we show results based on two models (OPE and OME-f) which predict quite different Γ_n/Γ_p ratios. Unfortunately, the dependence of the neutron spectra on variations of Γ_n/Γ_p is very weak (the same is true also for the proton spectra) and a precise extraction of the ratio from the KEK–E369 distribution is not possible. The problem of the small sensitivity

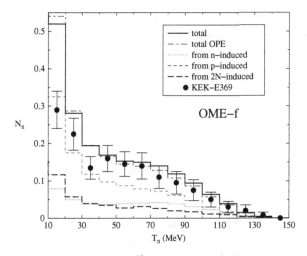

Fig. 10. Single-neutron spectrum for $^{12}_\Lambda$C calculated using two different weak interaction models, a π-exchange (OPE) model and a $\pi + \varrho + K + K^* + \eta + \omega$-exchange ($OME$-$f$) model, which predict quite different Γ_n/Γ_p ratios. The observed spectrum in the KEK–E369 experiment [63] is also shown for comparison

of N_n and N_p to variations of Γ_n/Γ_p can be overcome if one concentrates on another single–nucleon observable. The ratio Γ_n/Γ_p is defined in terms of the ratio between the number of primary weak decay neutrons and protons, $N_n^{\rm wd}$ and $N_p^{\rm wd}$,

$$\frac{\Gamma_n}{\Gamma_p} \equiv \frac{1}{2}\left(\frac{N_n^{\rm wd}}{N_p^{\rm wd}} - 1\right) \neq \frac{1}{2}\left(\frac{N_n}{N_p} - 1\right) \equiv R_1\left[\Delta T_n, \Delta T_p, \Gamma_2\right] , \qquad (50)$$

where the inequality holds due to two–body induced decays and (especially) nucleon FSI. This is valid in a situation, such as the experimental one, in which particular intervals of variability of the neutron and proton kinetic energy, ΔT_n and ΔT_p, are employed in the determination of N_n and N_p. This is more evident from Table 6, in which the function R_1 is given for $^5_\Lambda$He and $^{12}_\Lambda$C, for different nucleon energy thresholds $T_N^{\rm th}$ and for the OPE and OME-f models. For a given energy threshold, R_1 is closer to Γ_n/Γ_p for $^5_\Lambda$He than for $^{12}_\Lambda$C since FSI are larger in carbon. The ratio N_n/N_p (or R_1) is more sensitive to variations of Γ_n/Γ_p (see the differences between the OPE and OME-f calculations of Table 6) than N_n and N_p separately. Moreover, N_n/N_p is less affected by FSI than N_n and N_p. Therefore, measurements of N_n/N_p should permit to determine Γ_n/Γ_p with better precision.

Let us note that the nucleons originating from n– and p–induced processes are added *incoherently* (i.e., classically) in the intranuclear cascade calculation. However, for particular kinematics of the detected nucleons (for instance at low kinetic energies), an in principle possible quantum–mechanical interference effect between n– and p–induced channels should inevitably affect the

Table 6. Predictions for the quantity R_1 of (50) for $^5_\Lambda$He and $^{12}_\Lambda$C corresponding to different nucleon thresholds T_N^{th} and to the OPE, and OME-f models

| | | T_N^{th} (MeV) | | | |
		0	30	60	Γ_n/Γ_p
$^5_\Lambda$He	OPE	0.04	0.13	0.16	0.09
	OME-f	0.19	0.40	0.49	0.46
$^{12}_\Lambda$C	OPE	−0.06	−0.01	0.05	0.08
	OME-f	−0.01	0.09	0.21	0.34

observed distributions. Therefore, extracting the ratio Γ_n/Γ_p from experimental data with the help of a classical intranuclear cascade calculation may not be a clean task. To clarify better the issue, let us consider for instance the experimental measure of *single-proton* kinetic energy spectra. The relevant quantity is then the number of outgoing protons observed as a function of the kinetic energy T_p. Schematically, this observable can be written as:

$$N_p(T_p) \propto \left| \langle p(T_p)| \hat{O}_{FSI}\,\hat{O}_{WD}|\Psi_H\rangle \right|^2$$

$$= \left| \alpha \langle p(T_p)|\hat{O}_{FSI}|nn, \Psi_R\rangle + \beta \langle p(T_p)| \hat{O}_{FSI}|np, \Psi_{R'}\rangle \right|^2 , \quad (51)$$

where $|p(T_p)\rangle$ represents a many-nucleon final state with a proton whose kinetic energy is T_p. Moreover, in (51) the action of the weak decay operator $\hat{O}_{WD} \equiv \hat{O}_{\Lambda n \to nn} + \hat{O}_{\Lambda p \to np}$ produced the superposition:

$$\hat{O}_{WD}|\Psi_H\rangle = \alpha\,|nn, \Psi_R\rangle + \beta\,|np, \Psi_{R'}\rangle .$$

Here $|nn, \Psi_R\rangle$ ($|np, \Psi_{R'}\rangle$) is a state with a nn (np) primary pair moving inside a residual nucleus Ψ_R ($\Psi_{R'}$). Note that in the present schematic picture: $\Gamma_n \propto |\alpha|^2$ and $\Gamma_p \propto |\beta|^2$. Since both transition amplitudes entering the last equality of (51) are in general non-vanishing, interference terms between n- and p-induced decays are expected to contribute to $N_p(T_p)$. An amplitude $\langle p(T_p)|\hat{O}_{FSI}|nn, \Psi_R\rangle$ different from zero means that, due to nucleon final state interactions, a secondary proton has a non-vanishing probability to emerge from the nucleus with kinetic energy T_p even if the weak process was n-induced (i.e., without primary protons). While for high kinetic energies this amplitude is expected to be almost vanishing, as long as T_p decreases its contribution could produce an important interference effect.

An interference-free observation would imply the measurement of all the quantum numbers of the final nucleons and residual nucleus. While this is an impossible experiment, what is certain is that the magnitude of the interference can be reduced if one measures in a more accurate way the final state.

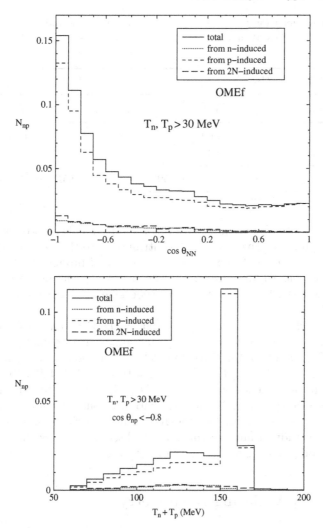

Fig. 11. Number of neutron-proton pairs as a function of the cosine of the relative angle between the two nucleons (*upper panel*) and as a function of the sum of kinetic energies (*lower panel*). Shown plots are for $^{12}_{\Lambda}$C

For this reason, two-nucleon coincidence observables are expected to be less affected by interferences than single-nucleon ones and thus more reliable for determining Γ_n/Γ_p. Upper panel of Fig. 11 shows the np pair opening angle distribution in the case of $^{12}_{\Lambda}$C. The total spectrum N_{np} has been decomposed into the components $N_{np}^{\Lambda n \rightarrow nn}$, $N_{np}^{\Lambda p \rightarrow np}$ and $N_{np}^{\Lambda np \rightarrow nnp}$. A nucleon energy threshold of 30 MeV has been used in the calculation. Lower panel of Fig. 11 corresponds to the kinetic energy correlation of np pairs: it is again for $^{12}_{\Lambda}$C and $T_N^{th} = 30$ MeV, but now only back-to-back angles ($\cos\theta_{np} \leq -0.8$) have been

Table 7. Predictions for the ratio $R_2 \equiv N_{nn}/N_{np}$ for $^5_\Lambda$He and $^{12}_\Lambda$C. An energy threshold of $T_N^{th} = 30\,\mathrm{MeV}$ and a pair opening angle of $\cos\theta_{NN} \leq -0.8$ have been considered

	$^5_\Lambda$He		$^{12}_\Lambda$C	
	N_{nn}/N_{np}	Γ_n/Γ_p	N_{nn}/N_{np}	Γ_n/Γ_p
OPE	0.25	0.09	0.24	0.08
OME-f	0.61	0.46	0.43	0.34
EXP	$0.45 \pm 0.11 \pm 0.03$ [50]		$0.51 \pm 0.13 \pm 0.05$ [51]	

taken into account. We note how both the n–induced and the two–nucleon induced decay processes give very small contributions to the total distributions in Fig. 11. Nevertheless, these decay processes could produce non–negligible interference terms. To minimize this effect, one could consider, for instance, not only back-to-back angles but also nucleon kinetic energies in the interval $150 \div 170\,\mathrm{MeV}$.

In Table 7 the ratio N_{nn}/N_{np} predicted by the OPE and OME-f models for $^5_\Lambda$He and $^{12}_\Lambda$C is given for $\cos\theta_{np} \leq -0.8$ and a nucleon energy threshold of $30\,\mathrm{MeV}$. The results of the OME-f model are in reasonable agreement with the 2004 KEK datum for $^5_\Lambda$He [50], $\Gamma_n/\Gamma_p = 0.45 \pm 0.11 \pm 0.03$. Preliminary analysis on the dependence of N_{nn}/N_{np} on Γ_n/Γ_p and Γ_{2N} [13], give neutron-to-proton ratios rather small if compared with previous determinations, which gave values $\gtrsim 1$. Although further (theoretical and experimental) confirmation is needed, the study of nucleon coincidence observables offers the possibility to solve the longstanding puzzle on the Γ_n/Γ_p ratio.

7 The Parity-Violating Asymmetry

The study of the decay of polarized hypernuclei provides complementary information about the ΛN interaction, such as the spin-parity structure of the weak process, or the magnetic moments of hypernuclei. When the polarized hypernucleus is created, and due to the interference between the PC and PV amplitudes, the distribution of the emitted protons in the weak decay displays an angular asymmetry with respect to the polarization axis. A complete derivation of the expressions for the evaluation of the asymmetry parameter can be found in [65], where the starting point is given by the intensity of outgoing nucleons:

$$I(\chi) = \mathrm{Tr}\left(\mathcal{M}\hat{\rho}\mathcal{M}^\dagger\right)$$
$$= \sum_{FM_IM_I'} \langle F \mid \mathcal{M} \mid M_I\rangle\langle M_I \mid \hat{\rho} \mid M_I'\rangle\langle M_I' \mid \mathcal{M}^\dagger \mid F\rangle, \qquad (52)$$

where χ is the angle between the direction of the proton and the polarization axis and $\hat{\rho}$ represents the density matrix for the polarized J-spin hypernucleus (see Fig. 12). For pure vector polarization perpendicular to the plane of the (π^+, K^+) reaction, the density matrix is given by the expression:

$$\hat{\rho}(J) = \frac{1}{2J+1} \left(1 + \frac{3}{J+1} P_y S_y \right) , \tag{53}$$

with S_y being the J-spin operator along the polarization axis and P_y the hypernuclear polarization created in the production reaction. Introducing (53) in the expression of $I(\chi)$ one obtains:

$$I(\chi) = I_0 \left(1 + \frac{3}{J+1} P_y \frac{\text{Tr}\left(\mathcal{M} S_y \mathcal{M}^\dagger \right)}{\text{Tr}\left(\mathcal{M} \mathcal{M}^\dagger \right)} \right) = I_0 \left(1 + \mathcal{A} \right) , \tag{54}$$

where I_0 is the isotropic intensity for the unpolarized hypernucleus,

$$I_0 = \frac{\text{Tr}(\mathcal{M}\mathcal{M}^\dagger)}{2J+1} , \tag{55}$$

and $\mathcal{A}$ the asymmetry.

In [65] it is shown that, for pure vector polarization, $\mathcal{A} = P_y A_p \cos\chi$, where A_p is characteristic of the weak decay mechanism and comes from the interference of the PC and PV amplitudes. Its value is found theoretically by evaluating

$$A_p = \frac{3}{J+1} \frac{\sum_{M_I} \sigma(M_I) M_I}{\sum_{M_I} \sigma(M_I)} , \tag{56}$$

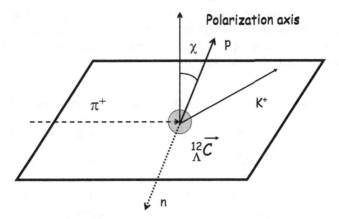

Fig. 12. Schematic illustration of a (π^+, K^+) reaction on ^{12}C. The typical kinematical conditions of such reaction, with a pion momentum of $\sim 1.05\,\text{GeV}$ and small kaon angles $2° \le \theta_K \le 15°$, produces $^{12}_\Lambda$C hypernuclei with large spin polarization aligned preferentially along the axis normal to the reaction plane. The PV asymmetry is obtained by looking at the angular distribution of the weak decay protons

with $\sigma(M_I)$ the intensity of protons exiting along the quantization axis z ($\hat{k}_1$) for a spin projection M_I of the hypernucleus. At $\chi = 0°$ the asymmetry in the distribution of protons is thus determined by the product $P_y A_p$. Therefore, extracting from experiments A_p involves the theoretical (model dependent) evaluation of the hypernuclear polarization.

In the weak coupling scheme, simple angular momentum algebra relations relate the hypernuclear polarization to the Λ polarization

$$p_\Lambda = \begin{cases} -\dfrac{J}{J+1}P_y & \text{if } J = J_C - \frac{1}{2} \\ P_y & \text{if } J = J_C + \frac{1}{2} \end{cases}, \qquad (57)$$

where J_C is the spin of the nuclear core. Following the same scheme, it is convenient to introduce the intrinsic Λ asymmetry parameter

$$\alpha_\Lambda = \begin{cases} -\dfrac{J+1}{J}A_p & \text{if } J = J_C - \frac{1}{2} \\ A_p & \text{if } J = J_C + \frac{1}{2} \end{cases}, \qquad (58)$$

such that $P_y A_p = p_\Lambda \alpha_\Lambda$, which becomes characteristic of the elementary Λ decay process, $\overrightarrow{\Lambda}N{\rightarrow}NN$, taking place in the nuclear medium and in principle, independent of the hypernuclear size.

For s-shell hypernuclei one can write a simplified expression in terms of the elementary transitions appearing in Table 8,

$$a_\Lambda = \frac{2\sqrt{3}\,\text{Re}[\,a\,e^* - b\,(\,c - \sqrt{2}\,d\,)^*/\sqrt{3} + f(\sqrt{2}\,c + d\,)^*]}{|a|^2 + |b|^2 + 2\,[\,|c|^2 + |d|^2 + |e|^2 + |f|^2]}. \qquad (59)$$

The comparison of theoretical models with data is difficult and disappointing. While theory predicts negative and large values of a_Λ for both s- and p-shell hypernuclei, experiments give different sign for the asymmetry of $^5_\Lambda\overrightarrow{\text{He}}$ and $^{12}_\Lambda\overrightarrow{\text{C}}$, although some of the measurements contain large uncertainties, giving values compatible with zero.

Table 8. Allowed weak transitions when the initial system is an s-shell hypernucleus

	ΛN $^{2S+1}L_J$	NN $^{2S'+1}L'_J$	NN isospin	PC/PV	operator	size
a:	$^1S_0$	$^1S_0$	1	PC	$\hat{1}, \sigma_1\sigma_2$	1
b:		$^3P_0$	1	PV	$(\sigma_1 - \sigma_2)q, (\sigma_1 \times \sigma_2)q$	q/M_N
c:	$^3S_1$	$^3S_1$	0	PC	$\hat{1}, \sigma_1\sigma_2$	1
d:		$^3D_1$	0	PC	$(\sigma_1 \times q)(\sigma_2 \times q)$	q^2/M_N^2
e:		$^1P_1$	0	PV	$(\sigma_1 - \sigma_2)q, (\sigma_1 \times \sigma_2)q$	q/M_N
f:		$^3P_1$	1	PV	$(\sigma_1 + \sigma_2)q$	q/M_N

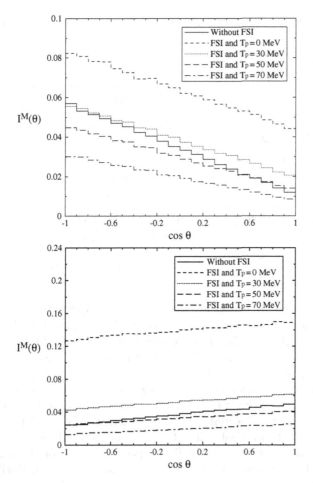

Fig. 13. Angular intensity of protons emitted per NMWD for $^5_\Lambda\overrightarrow{\text{He}}$ (*top*) and $^{12}_\Lambda\overrightarrow{\text{C}}$ (*bottom*)

Figure 13 shows the proton intensity obtained by [66] for the non-mesonic decay of $^5_\Lambda\overrightarrow{\text{He}}$ and $^{12}_\Lambda\overrightarrow{\text{C}}$ using the full one-meson-exchange model with the NSC97f potential. Note that the hypernuclear polarization has been taken to be $P_y = 1$ (i.e. $p_\Lambda = 1$ for $^5_\Lambda\overrightarrow{\text{He}}$ and $-1/2$ for $^{12}_\Lambda\overrightarrow{\text{C}}$) so that the asymmetry parameter can be directly extracted from the values of the intensity at $\theta = 0°$ and $\theta = 180°$ through the relation $a_\Lambda^{\text{M}} = \dfrac{1}{p_\Lambda}\dfrac{I^{\text{M}}(0°) - I^{\text{M}}(180°)}{I^{\text{M}}(0°) + I^{\text{M}}(180°)}$. Note that in writing the previous expression, one assumes that the experimental proton intensity $I^{\text{M}}(\theta)$ has the same θ-dependence as the intensity for primary protons, $I^{\text{M}}(\theta) = I_0^{\text{M}}[1 + p_\Lambda\, a_\Lambda^{\text{M}}\cos\theta]$. The continuous histograms correspond to the intensity $I(\theta)$ of primary protons. The inclusion of the nucleon FSI strongly modifies the spectra. With vanishing kinetic energy detection

Table 9. Measured asymmetry parameter for the non–mesonic weak decay of $^5_\Lambda \overrightarrow{\text{He}}$ and $^{12}_\Lambda \overrightarrow{\text{C}}$

Model	a^M_Λ ($^5_\Lambda$He)	a^M_Λ ($^{12}_\Lambda$C)
OPE	−0.25	−0.34
OME	−0.68	−0.73
FSI and $T^{\text{th}}_\text{p} = 0\,\text{MeV}$	−0.30	−0.16
FSI and $T^{\text{th}}_\text{p} = 30\,\text{MeV}$	−0.46	−0.37
FSI and $T^{\text{th}}_\text{p} = 50\,\text{MeV}$	−0.52	−0.51
FSI and $T^{\text{th}}_\text{p} = 70\,\text{MeV}$	−0.55	−0.65
KEK–E462 (80 MeV) [67]	$0.09 \pm 0.14 \pm 0.04$ (inclusive)	
	0.31 ± 0.22 (np in coincidence)	
KEK–E508 (preliminary)[67]		-0.44 ± 0.32

threshold, T^{th}_p, the intensities are strongly enhanced, especially for $^{12}_\Lambda \overrightarrow{\text{C}}$. For $T^{\text{th}}_\text{p} = 30$ or $50\,\text{MeV}$, the spectra are closer to $I(\theta)$, although with a different slope, reflecting the fact that FSI are responsible for a substantial fraction of outgoing protons with energy below these thresholds. A further reduction of $I^M(\theta)$ is observed for $T^{\text{th}}_\text{p} = 70\,\text{MeV}$.

Looking at the results shown in Table 9, it is evident that the OME results are in agreement with the $^{12}_\Lambda \overrightarrow{\text{C}}$ datum but inconsistent with the $^5_\Lambda \overrightarrow{\text{He}}$ one. One also sees that the OPE asymmetries are systematically smaller, though less realistic from the theoretical point of view, than the OME ones. The analysis made in [66] proves that only small and positive values of the primary a_Λ, not predicted by any existing model, could reduce the measured a^M_Λ to small and positive values, compatible with the experimental information on $^5_\Lambda \overrightarrow{\text{He}}$. In order to better establishing the sign and magnitude of a^M_Λ for s- and p-shell hypernuclei, new and/or improved experiments, will be important.

8 Effective Field Theory Approach

Although impressive progress has been achieved in hypernuclear physics through the use of phenomenological models, there is a lack of understanding of the underlying physics on more fundamental grounds. We have seen that the one-pion-exchange mechanism, which has a typical range of $m_\pi^{-1} \sim 1.4$ fm, has proven to be very efficient in describing the long range part of the $|\Delta S| = 1$ ΛN interaction, and that, in order to account for shorter distances, either more massive particles are exchanged, or an effective quark Hamiltonian is used. Those model dependencies reveal large uncertainties in the physical origin of the short-distance components, which in turn influence low-momentum physics.

In the non strange sector, Effective Field Theories (EFT) have been successfully used to deal with strong interaction physics in the non perturbative regime, specially in the two- and three-nucleons sector [68]. The main idea behind an EFT is that the physics governing the low-energy regime should not depend on the detailed knowledge of the physics governing the high energy regime. Therefore, one can describe the interaction of two particles at low energy as a finite series of terms of increasing dimension, where the high energy physics has been integrated out and, in practice, has been encoded in the coefficients of such expansion, the *low energy coefficients*, LECs. If the problem is suited for an EFT approach, the series has to be finite and indeed, convergent.

Effective field theories are standard techniques in nuclear physics used to systematically approach physical processes where one can identify different and well separated scales. In the weak $|\Delta S| = 1$ interaction one can easily identify three scales: the baryon masses, with a typical value of $\overline{M} = (M_N + M_\Lambda)/2 \approx 1027$ MeV, the pion mass, $m_\pi \approx 138$ MeV, and the typical value of the momentum of the final nucleons, $|\mathbf{p}| \approx 420$ MeV/c. Within the EFT scheme, high-momentum modes in the Lagrangian (with a mass $\gtrsim m_\varrho$), are replaced by contact operators of increasing dimension and compatible with the underlying physical symmetries. This formulation leads in principle to an infinite number of terms in the Lagrangian. Therefore, an expansion scheme is needed to truncate it and achieve a controlled and stable expansion. Within this expansion predictions are made for physical observables. The predictive power achieved in the $|\Delta S| = 0$ sector is remarkable and work in this direction for the weak ΛN has already started [41, 42] but is still insufficient. In contrast to the strong NN interaction, where one can study the very low energy regime, the $\Lambda N \to NN$ reaction responsible for the decay of hypernuclei produces nucleons with a kinetic energy of ≈ 80 MeV each, and a description based on contact terms is not realistic. Given the large energy release, it is necessary to include the pion ($m_\pi \approx 138$ MeV) and the kaon ($m_K \approx 494$ MeV) as dynamical fields. Note that, in principle, SU(3) would also suggest including the heavier pseudoscalar η meson ($m_\eta \approx 550$ MeV).

Another way to see what are the relevant explicit degrees of freedom to be included in our EFT is to look at a typical OME potential, the pion-exchange potential for instance,

$$V_{\text{ps}}^{\text{PC}} = -G_{\text{F}} m_\pi^2 \, \frac{g_{\text{NN}\pi}}{2 \, M_{\text{S}}} \, \frac{g_{\Lambda N\pi}}{2 \, M_{\text{W}}} \, \frac{q^2}{m_\pi^2} \, \frac{1}{1 + \left(\dfrac{q}{m_\pi}\right)^2} \, \boldsymbol{\sigma}_1 \hat{q} \, \boldsymbol{\sigma}_2 \hat{q} \,, \qquad (60)$$

where I only quoted the PC piece and omitted the isospin dependence for simplicity. Expanding this expression in powers of the exchanged momentum

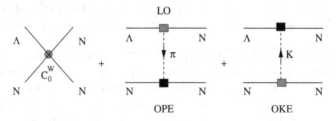

Fig. 14. Lowest order contribution to the weak $\Lambda N \rightarrow NN$ diagram. *Shaded vertices* represent weak vertices while *solid ones* represent strong vertices. A circle stands for a contact non derivative operator and a square for an insertion of a derivative operator

q, one can identify the relevant quantity in the low energy regime to be: $\dfrac{g_{NN\pi}}{2M_S} \dfrac{g_{\Lambda N\pi}}{2M_W} \dfrac{1}{m_\pi^2}$, which depends in the numerator on the baryon-baryon-pion couplings (strong and weak), and it is suppressed by the inverse of the square of the meson mass, giving us an estimate of what are the relative contributions of the different mesons in the different ranges of the interaction.[6]

In the approach followed by [42], the leading order EFT gets contributions from the diagrams depicted in Fig. 14. Notice that this approach is equivalent to a chiral expansion of the vertices entering the $\Lambda N \rightarrow NN$ transition, while using a phenomenological approach to account for the strong interaction between the baryons involved in the process. Those vertices are nothing else but combinations of the five Dirac bilinear covariants. Their relativistic form encodes all the orders in a momentum expansion, therefore, their chiral expansion would better allow the power counting by comparing non relativistic terms of size 1, p/M, etc. In order to avoid formal inconsistencies from the chiral point of view, it is better to directly rely on the terms which enter at each order given by the symmetries of the physics problem. To illustrate this, suppose that we have the ΛN pair in a $L = 0$ state. The proper way to parametrize the contact terms is in the form of the famous, old a, b, c, d, e, f coefficients of [8], shown in Table 8. The a and c transitions can only be produced by combining the $\hat{1} \cdot \delta(r)$ and $\boldsymbol{\sigma}_1\boldsymbol{\sigma}_2 \cdot \delta(r)$ operators, where $\delta(r)$ stands for the contact interaction, while in order to produce d, one needs a second order operator, $(\boldsymbol{\sigma}_1 \times \boldsymbol{q})(\boldsymbol{\sigma}_2 \times \boldsymbol{q})$. The PV f transition can only be produced by the combination of two spin conserving operators $(\boldsymbol{\sigma}_1 + \boldsymbol{\sigma}_2)\{\boldsymbol{p}_1 - \boldsymbol{p}_2, \delta(r)\}$ and $(\boldsymbol{\sigma}_1 + \boldsymbol{\sigma}_2)[\boldsymbol{p}_1 - \boldsymbol{p}_2, \delta(r)]$, where $\{ \ , \ \}$ denotes an anti-commutator, $[\ , \ ]$ a commutator and $\boldsymbol{p}_i$ is the derivative operator[7]. Finally, the PV b and e transitions can

---

[6] The ratio of this quantity for π, η and K for the $\Lambda n \rightarrow nn$ process comes out to be: $4.94 \div 0.25 \div 1.11$ when the NSC97f [53] strong interaction model is used.

[7] We are assuming that $\boldsymbol{p}_1 - \boldsymbol{p}_2$ is small enough to disregard higher powers of the derivative operators $\boldsymbol{p}_1 - \boldsymbol{p}_2$.

only be produced by: $(\boldsymbol{\sigma}_1 - \boldsymbol{\sigma}_2)\{\boldsymbol{p}_1 - \boldsymbol{p}_2\,,\,\delta(\boldsymbol{r})\}$, $(\boldsymbol{\sigma}_1 - \boldsymbol{\sigma}_2)[\boldsymbol{p}_1 - \boldsymbol{p}_2\,,\,\delta(\boldsymbol{r})]$, $\mathrm{i}\,(\boldsymbol{\sigma}_1 \times \boldsymbol{\sigma}_2)\{\boldsymbol{p}_1 - \boldsymbol{p}_2\,,\,\delta(\boldsymbol{r})\}$, and $\mathrm{i}\,(\boldsymbol{\sigma}_1 \times \boldsymbol{\sigma}_2)[\boldsymbol{p}_1 - \boldsymbol{p}_2\,,\,\delta(\boldsymbol{r})]$.

Rearranging all the terms[8] one can write the most general Lorentz invariant potential, with no derivatives in the fields, for the four-fermion interaction in momentum space up to $\mathcal{O}(q^2/M^2)$ (in units of $G_{\mathrm{F}} = 1.166 \times 10^{-11}\,\mathrm{MeV}^{-2}$):

$$
\begin{aligned}
V_{4\mathrm{P}}(\boldsymbol{q}) \;=\;& C_0^0 + C_0^1\,\boldsymbol{\sigma}_1 \cdot \boldsymbol{\sigma}_2 \\[4pt]
& + C_1^0\,\frac{\boldsymbol{\sigma}_1 \cdot \boldsymbol{q}}{2\,\overline{M}} + C_1^1\,\frac{\boldsymbol{\sigma}_2 \cdot \boldsymbol{q}}{2\,M} + \mathrm{i}\,C_1^2\,\frac{(\boldsymbol{\sigma}_1 \times \boldsymbol{\sigma}_2)\cdot\boldsymbol{q}}{2\,\tilde{M}} \\[4pt]
& + C_2^0\,\frac{\boldsymbol{\sigma}_1 \cdot \boldsymbol{q}\,\boldsymbol{\sigma}_2 \cdot \boldsymbol{q}}{4M\overline{M}} + C_2^1\,\frac{\boldsymbol{\sigma}_1 \cdot \boldsymbol{\sigma}_2\,q^2}{4M\overline{M}} + C_2^2\,\frac{q^2}{4M\tilde{M}}\,,
\end{aligned}
\tag{61}
$$

where $\overline{M} = (M + M_\Lambda)/2$ and $\tilde{M} = (3M + M_\Lambda)/4$. C_i^j is the jth low energy coefficient at ith order. The approach of [42] includes a regularizing form factor at each vertex of the OPE and OKE diagrams, of the same type as discussed in previous sections. Note that in principle, form factor effects would be generated term by term in the chiral expansion through a higher-order chiral loop. Nevertheless, the present prescription is more phenomenological, where the form factor has been designed to give the correct physics in a particular energy-momentum region. This approach can result in inaccuracies when applied to other regions, but this is an effective way to codify phenomena for which one cannot explain in detail their origin [69]. From the former derivation, it is clear that the form of the contact terms is model independent. The LECs represent the short distance contributions and their size depends on how the theory is formulated, and more specifically upon the chiral order we are working. With respect to the isospin part of the 4-fermion (4P) interaction, we should *in principle* allow for both, $\Delta I = 1/2$ and $\Delta I = 3/2$ transitions. Matrix elements of the $\Delta I = 1/2$ ($\Delta I = 3/2$) operator can be easily included by assuming the Λ to behave like an isospin $|\,1/2\ -1/2\,\rangle$ ($|\,3/2\ -1/2\,\rangle$) state and introducing an isospin dependence in the $\Delta I = 1/2$ ($\Delta I = 3/2$) transition potential of the type $\boldsymbol{\tau} \cdot \boldsymbol{\tau}$ ($\boldsymbol{\tau}_{3/2} \cdot \boldsymbol{\tau}$), where $\boldsymbol{\tau}$ ($\boldsymbol{\tau}_{3/2}$) is the $1/2 \to 1/2$ ($1/2 \to 3/2$) isospin transition operator. The spherical components of the $\Delta I = 1/2$ and $\Delta I = 3/2$ operators have the matrix elements [40]:

$$
\langle\,1/2\ m'\,|\,\tau_{1/2}^{(i)}\,|\,1/2\ m\,\rangle = \langle\,1/2\ m\ 1\ i\,|\,1/2\ m'\,\rangle \qquad i = \pm 1, 0\,,
\tag{62}
$$

$$
\langle\,3/2\ m'\,|\,\tau_{3/2}^{(i)}\,|\,1/2\ m\,\rangle = \langle\,1/2\ m\ 1\ i\,|\,3/2\ m'\,\rangle \qquad i = \pm 1, 0\,.
\tag{63}
$$

Since the 4P potential in configuration space is obtained by Fourier transforming $V_{4\mathrm{P}}(\boldsymbol{q})$, one has to smear the resulting delta functions. Typically,

---

[8] In order to derive the potential, one assumes that, since the two interacting particles in the initial state are bound in a hypernucleus, one can neglect the part in which $\boldsymbol{p}_1 - \boldsymbol{p}_2$ acts on the ΛN state, compared to $\boldsymbol{p}_1 - \boldsymbol{p}_2$ acting on the final NN state. This amounts to neglecting terms containing the relative initial $\boldsymbol{k}_{\mathrm{i}}$ momentum in the expansion, and at the same time to approximate the relative final $\boldsymbol{k}_{\mathrm{f}}$ momentum by the momentum transfer $-\boldsymbol{q}$.

one uses a normalized Gaussian form for the 4-fermion contact potential, $f_{ct}(r) = \exp(-r^2/\delta^2)/(\delta^3 \pi^{3/2})$, where δ is taken to be of the order of the range given by the first meson excluded in our formalism (as compared to a one-meson-exchange model for instance), the ϱ meson range, $\delta = \sqrt{2}\, m_\varrho^{-1} \approx 0.36$ fm. The dependence of our LECs on the hard cut-off δ should be explored, since for the EFT to be applicable, these LEC should be not only of natural size, given the dimensions of the problem, but also quite stable as higher order contributions are included. The expression of the $V_{4P}(r)$ potential up to NLO PC terms is then:[9]

$$
\begin{aligned}
V_{4P}(\boldsymbol{r}, \boldsymbol{\tau}) = \{ &\, C_0^0 + C_0^1\, \boldsymbol{\sigma}_1 \boldsymbol{\sigma}_2 \\
&+ \frac{2r}{\delta^2} \left[C_1^0\, \frac{\boldsymbol{\sigma}_1 \hat{r}}{2\,\overline{M}} + C_1^1\, \frac{\boldsymbol{\sigma}_2 \hat{r}}{2\,M} + C_1^2\, \frac{(\boldsymbol{\sigma}_1 \times \boldsymbol{\sigma}_2)\, \hat{r}}{2\,\tilde{M}} \right] \\
&+ \frac{1}{\delta^2} \left\{ \left[6 - \left(\frac{2r}{\delta}\right)^2 \right] \left[\frac{C_2^2}{4\tilde{M}M} + \frac{\boldsymbol{\sigma}_1 \boldsymbol{\sigma}_2}{4\tilde{M}M} \left(\frac{C_2^0}{3} + C_2^1 \right) \right] \right. \\
&\left. - \frac{1}{3}\, \frac{C_2^0}{4\tilde{M}M}\, S_{12}(\hat{r})\, \frac{4r^2}{\delta^2} \right\} \\
&\times\ f_{ct}(r) \\
&\times\ \left(C_s\, \hat{1} + C_v\, \boldsymbol{\tau}_1 \cdot \boldsymbol{\tau}_2 + C_{3/2}\, \boldsymbol{\tau}_{3/2} \cdot \boldsymbol{\tau}_2 \right) \,.
\end{aligned} \tag{64}
$$

In order to obtain the values of the LECs at a given order, one should fit a set of experimental data. One of the most severe restrictions of this approach is the lack of a quantitative reliable and independent set of experimental numbers, which limits the order up to which one can perform the analysis. In hypernuclear decay one can identify three independent observables, the proton-induced and neutron-induced rates Γ_p and Γ_n (or the total non-mesonic decay rate Γ_{nm} and the neutron-to-proton ratio), and the PV asymmetry $\mathcal{A}$. Besides, observables from different hypernuclei can be related through hypernuclear structure coefficients. Therefore, one does not expect measurements from different p-shell hypernuclei, say, A = 12 and 16, to provide different constraints, but we expect so when including data from s-shell hypernuclei, like A = 5. In the work of [42], not all the available measurements were used in the fitting strategy, but only the more recent measurements from the last 14 years were used. The authors excluded those recent data of the ratio Γ_n/Γ_p whose central values were larger than 1, and whose error bars were larger than 100%, as well as old values inconsistent with the more recent results of improved experiments. The results of [42] are summarized in Table 10, where we show the effects of including on top of the weak OPE

---

[9] Matrix elements of the first and second PV terms in (64) can be related in a coupled spin basis formalism taking into account that $C_1^0\, \dfrac{\boldsymbol{\sigma}_1 \hat{r}}{2\overline{M}} + C_1^1\, \dfrac{\boldsymbol{\sigma}_2 \hat{r}}{2M} \approx$ $\dfrac{\boldsymbol{\sigma}_2 \hat{r}}{2\tilde{M}}\, \{ C_1^0\, (-1)^{(1-\delta_{SS_0})} + C_1^1 \}$, where S_0 and S are the spins of the initial ΛN pair and the final NN pair respectively, as in Sect. 4.2.

Table 10. Results obtained for the weak decay observables, when a fit to the Γ and $\frac{\Gamma_n}{\Gamma_p}$ for $^5_\Lambda$He, $^{11}_\Lambda$B and $^{12}_\Lambda$C is performed. The values in parentheses include α_Λ ($^5_\Lambda$He) in the fit

	π	$+ K$	$+ LO$	$+ NLO$	EXP
$\Gamma\,(^5_\Lambda\text{He})$	0.42	0.23	0.43	0.44 (0.44)	0.41 ± 0.14 [26]
					0.50 ± 0.07 [46]
$\frac{\Gamma_n}{\Gamma_p}\,(^5_\Lambda\text{He})$	0.08	0.50	0.56	0.55 (0.55)	0.93 ± 0.55 [26]
					0.50 ± 0.10 [70]
$\alpha_\Lambda\,(^5_\Lambda\text{He})$	-0.25	-0.60	-0.80	0.28 (0.24)	0.24 ± 0.22 [71]
$\Gamma\,(^{11}_\Lambda\text{B})$	0.63	0.36	0.87	0.88 (0.88)	0.95 ± 0.14 [46]
$\frac{\Gamma_n}{\Gamma_p}\,(^{11}_\Lambda\text{B})$	0.10	0.43	0.84	0.92 (0.92)	$1.04^{+0.59}_{-0.48}$ [26]
$\mathcal{A}\,(^{11}_\Lambda\text{B})$	-0.09	-0.22	-0.22	0.09 (0.08)	-0.20 ± 0.10 [64]
$\Gamma\,(^{12}_\Lambda\text{C})$	0.75	0.41	0.95	0.93 (0.93)	1.14 ± 0.20 [26]
					0.89 ± 0.15 [46]
					0.83 ± 0.11 [48]
$\frac{\Gamma_n}{\Gamma_p}\,(^{12}_\Lambda\text{C})$	0.08	0.35	0.67	0.77 (0.77)	0.87 ± 0.23 [72]
$\mathcal{A}\,(^{12}_\Lambda\text{C})$	-0.03	-0.06	-0.05	0.03 (0.02)	-0.01 ± 0.10 [64]
$\hat{\chi}^2$			0.93	1.54 (1.15)	

mechanism, the kaon-exchange contribution and the leading order and next-to-leading order contact terms. Obviously, no parameters were fitted for the results shown in the first two columns, since all the unknown constants were obtained by assuming flavor-symmetry. These results can be compared to the numbers shown in previous sections. The third column represents the leading order contribution, which includes contact terms of size unity. These contact terms are PC operators and they contribute with four free parameters, C_0^0, C_0^1, C_s and C_v, which are fitted to reproduce the total and partial decay rates for all three hypernuclei. Their inclusion is enough to restore the total decay rate, which is now in agreement with experiment for the three nuclei. The impact on the ratio is noteworthy: the value for $^5_\Lambda$He increases by 10% while the Γ_n/Γ_p ratios for $^{11}_\Lambda$B and $^{12}_\Lambda$C almost double due to the appearance of new partial waves. This is an example of the different impact certain operators can have for s- and p-shell hypernuclei. The effect on the asymmetry is opposite, almost no change for $A = 11$ and 12, but a 30% change for $A = 5$, behavior that can be understood by following a similar argument as

Table 11. LEC coefficients corresponding to the LO calculation. The values in parentheses include α_Λ ($^5_\Lambda$He) in the fit

	+ LO PC	+ LO PC + LO PV
C_0^0	-1.54 ± 0.39	-1.31 ± 0.41 (-1.04 ± 0.33)
C_0^1	-0.87 ± 0.24	-0.70 ± 0.35 (-0.57 ± 0.27)
C_1^0	$- - -$	-5.82 ± 5.31 (-4.49 ± 1.57)
C_1^1	$- - -$	2.47 ± 3.13 $(\ 1.84 \pm 1.93)$
C_1^2	$- - -$	-5.68 ± 3.13 (-4.47 ± 2.31)
$C_{\rm s}$	5.01 ± 1.26	4.68 ± 0.67 $(\ 5.97 \pm 0.86)$
$C_{\rm v}$	1.45 ± 0.38	1.22 ± 0.20 $(\ 1.56 \pm 0.26)$

before.[10] The magnitudes of the four parameters, C_0^0, C_0^1, $C_{\rm s}$ and $C_{\rm v}$, listed in Table 11, are all around their natural size of unity, with the exception of $C_{\rm s}$ which is about a factor of five larger. Note the substantial error bars on all the parameters, reflecting the uncertainties in the measurements.

The next-to-leading terms introduce three new parameters, at order $q/M_{\rm N}$, which contribute with the coefficients C_1^0, C_1^1, and C_1^2. Surprisingly, in contrast to what one would expect for higher-order terms, the second column of Table 11 shows that the parameters for the PV contact terms are larger than the ones for the PC terms, but in fact, two of the three new constants are compatible with zero. One should note though that, the largest contact term still corresponds to an isoscalar, spin-independent central operator, which appears al LO. On the other hand, including the three new parameters does not substantially alter the previously fitted ones, which would support the validity of the expansion. Regarding their effect on the observables (see the fourth numerical column of Table 10), the PV contact terms barely modify the total decay rates. Neither the partial rates for helium. For boron and carbon, those partial rates are slightly modified, giving $\Gamma_{\rm n}/\Gamma_{\rm p}$ ratios 9–13% larger. The only observable which gets significantly affected is the asymmetry, as one should expect for an observable which results from the interference between PV and PC amplitudes. This observable changes sign for all three hypernuclei, moving the $^5_\Lambda$He value within the measured range at the expense of the one for $^{11}_\Lambda$B. This shift occurs without any asymmetry data constraining the fit. In order to further understand this behavior, [42] performed a number of fits including the asymmetry data points of either $^5_\Lambda$He or $^{11}_\Lambda$B or both, Tables 10 and 11 display (values in parenthesis) the result of one of those fits. They find that the two present experimental values for A = 5 and 11 cannot be fitted

---

[10] Note that for $^5_\Lambda$He we quote the value of the intrinsic Λ asymmetry parameter, a_Λ, which is experimentally accessible, while for p-shell hypernuclei the accessible quantity is $\mathcal{A}$.

simultaneously with this set of contact terms. Perhaps, future experiments will have to settle this issue.

Reference [42] also explores the dependence of the results on the Gaussian regulator and on the pion and kaon form factors. For values of δ ranging from 0.3 to 0.4 fm (from 900 to 500 MeV), the results for the total rate, the neutron-to-proton ratio and the $^5_\Lambda$He asymmetry are remarkably insensitive. This is not the case for the $^{11}_\Lambda$B and $^{12}_\Lambda$C asymmetries, the only predicted values in the table, which show a variation of 50% around its value at $\delta = 0.36$ fm. On the contrary, their results are insensitive to reasonable variations of the cut off in the OPE and OKE potentials. Allowing variations of the pion and kaon cut offs between 1500 and 2000 MeV, the results were exactly the same up to two digits for the observables, while the variations on the LECs values were hardly noticeable. Another possible source of model dependencies explored in [42] is the choice of the strong baryon-baryon interaction model. Employing NN wave functions that are obtained with either the Nijmegen NSC97f or the NSC97a model in the fit leaves the observables almost unchanged, with the exception of the asymmetry parameter, which can change up to 50%. The obtained couplings can easily absorb the changes but remain compatible within their error bars. An interesting point was made there regarding the relevance of $\Delta I = 3/2$ transitions. The inclusion of such operator did not help in constraining any of the low energy parameters, neither in improving the description of the weak decay process. The net effect was to shift strength from the isoscalar C_s contact term to the new $C_{3/2}$ one. The conclusion on this point was that one can get a good fit without including $\Delta I = 3/2$ transitions, i.e. without considering violations of the $\Delta I = 1/2$ rule.

The dominance of the LO isoscalar, spin-independent central operator, has motivated recent theoretical studies which included the σ meson explicitly in the weak meson-exchange mechanism. Reference [73] uses a pseudoscalar parametrization for the weak coupling at the $\Lambda N \sigma$ vertex, while a scalar coupling for the strong $NN\sigma$ vertex. The weak coupling constants are then adjusted to reproduce the central experimental values for the total decay rate and Γ_n/Γ_p ratio for $^5_\Lambda$He. They found that even though the σ meson contributes to bring theoretical predictions for such observables close to the experimental values for $^5_\Lambda$He and $^{12}_\Lambda$C, it is not enough to reproduce the more recent data for the asymmetry. Their results also corroborate the moderate dependence of the intrinsic asymmetry parameter on the considered hypernucleus, also found in other theoretical approaches. In [74] on the other hand, the authors add to the $\pi + K +$ Direct-Quark mechanism the exchange of a scalar-isoscalar meson. Their model succeeds in reproducing the experimental data for four- and five-body hypernuclei fairly well, in particular, the more recent data for the asymmetry parameter for $^5_\Lambda$He. They also stress the need of a direct measurement of the decay of $^4_\Lambda$H to establish the possible violation of the $\Delta I = 1/2$ rule.

9 Summary and Outlook

Hypernuclear decay studies can be a good source of information on the weak $|\Delta S| = 1$ ΛN interaction, specially when a close collaboration between theorists and experimentalists happens, collaboration which allows us to pursue measurements where medium effects and interferences can be quantified.

We have seen that simultaneous studies of inclusive and exclusive reactions make the mesonic decay mode a reliable scenario to distinguish between different pion-nucleus optical potentials. Regarding the non-mesonic decay mechanism, theoretical models have finally succeeded in predicting total decay rates and neutron-to-proton ratios in agreement with the more recent experimental data, although more work is needed to understand the present discrepancies found for the parity-violating asymmetry. In this direction, it has been pointed out the relevant role played by the inclusion of a scalar-isoscalar term in the weak transition potential. Although the results are promising, more effort has to be invested to fully understand its implications.

Effective field theory methods have already been applied to the decay mechanism, but a larger set of accurate and independent data is needed to better constrain the leading order four-baryon contributions to the weak Lagrangian. This is needed to get a model independent understanding of the weak mechanism. In practice, it is required to establish the connection between effective field theories and one-meson-exchange approaches of hypernuclear decay. Perhaps, the analysis of weak production mechanisms, such as the pn $\rightarrow$ pΛ performed on light targets and using polarized proton beams, will be able to provide us with some of the required data.

Another aspect that has to be further explored is how realistic hypernuclear wave functions modify present predictions for hypernuclear decay. At the two-particle level, a rigorous evaluation of the $\Lambda N - \Sigma N$ coupling in the initial ΛN wave function [75], which involves new Σ weak decay channels, is needed.

There are a few relevant problems I have not discussed in these lectures due to lack of time. One of them is the study of multi-strange systems, which provides very useful information through the analysis of new two-body decay channels, $\Lambda\Lambda \rightarrow \Lambda$n, $\Lambda\Lambda \rightarrow \Sigma^0$n and $\Lambda\Lambda \rightarrow \Sigma^-$p. Another avenue not discussed but one that holds great promise for obtaining complementary information on processes which are difficult to measure experimentally, is their simulation on the lattice. With the advent of new powerful supercomputers, lattice QCD simulations of the hyperon-nucleon interaction near threshold and of non-leptonic hyperon decays are presently undertaken [76].

Acknowledgments

I would like to start by thanking the organizers of this school, J. Adam, P. Bydžovský, J. Mareš, L. Majling and J. Novotný for inviting me to participate. I would like to thank also the School Secretary, R. Ortová, for her valuable help. Thanks to all the participants, for a very pleasant working and social atmospheres. Finally, thank you very much to the people who helped me in the preparation of these lectures, providing me with scientific material and technical support, specially to Àngels Ramos and Isaac Vidaña.

References

1. E. Oset and A. Ramos: Prog. Part. Nucl. Phys. **41**, 191 (1998)
2. W. M. Alberico and G. Garbarino: Phys. Rep. **369**, 1 (2002)
3. W. M. Alberico and G. Garbarino: Weak decay of hypernuclei, In: Proc. *Hadron Physics*, ed by T. Bressani, A. Filippi and U. Wiedner (Amsterdam, IOS Press, 2005) p. 125; nucl-th/0410059
4. M. Danysz and J. Pniewski: Phil. Mag. **44**, 348 (1953)
5. W. M. Alberico, A. De Pace, G. Garbarino, and A. Ramos: Phys. Rev. C **61**, 044314 (2000)
6. J. F. Donoghue, E. Golowich, and B. R. Holstein: In *Dynamics of the Standard Model*, ed by T. Ericson and P.V. Landshoff (Cambridge University Press, 1992)
7. J. Nieves and E. Oset: Phys. Rev. C **47**, 1478 (1993)
8. R. H. Dalitz and G. Rajasekharan: Phys. Lett. **1**, 58 (1962); M. M. Block and R. H. Dalitz: Phys. Rev. Lett. **11**, 96 (1963)
9. J. F. Dubach, G. B. Feldman, B. R. Holstein, and L. de la Torre: Ann. Phys. **249**, 146 (1996)
10. A. Parreño, A. Ramos, and C. Bennhold: Phys. Rev. C **56**, 339 (1997)
11. A. Parreño and A. Ramos: Phys. Rev. C **65**, 015204 (2002)
12. T. Inoue, K. Sasaki, and M. Oka: Nucl. Phys. A **670**, 301 (2000)
13. G. Garbarino, A. Parreño, and A. Ramos: Phys. Rev. Lett. **91**, 112501 (2003)
14. G. Garbarino, A. Parreño, and A. Ramos: Phys. Rev. C **69**, 054603 (2004)
15. E. Oset, P. Fernández de Córdoba, L. L. Salcedo, and R. Brockmann: Phys. Rep. **188**, 79 (1990)
16. A. Parreño, A. Ramos, and E. Oset: Phys. Rev. C **51**, 2477 (1995)
17. F. E. Close: *An Introduction to Quarks and Partons* (Academic Press, London, 1980)
18. E. Oset and L. L. Salcedo: Nucl. Phys. A **443**, 704 (1985)
19. P. Fernández de Córdoba and E. Oset: Nucl. Phys. A **528**, 736 (1991)
20. T. Motoba, H. Bandō, T. Fukuda, and J. Žofka: Nucl. Phys. A **534**, 597 (1991)
21. U. Straub, J. Nieves, A. Faessler, and E. Oset: Nucl. Phys. A **556**, 531 (1993)
22. K. Itonaga, T. Motoba, and H. Bandō: Z. Phys. A **330**, 209 (1988); Nucl. Phys. A **489**, 683 (1988)
23. T. Motoba: Nucl. Phys. A **547**, 115c (1992)
24. E. Oset, L. L. Salcedo, and Q. N. Usmani: Nucl. Phys. A **450**, 67c (1986)
25. I. Kumagai-Fuse, S. Okabe, and Y. Akaishi: Phys. Lett. B **345**, 386 (1995)
26. J. J. Szymanski et al: Phys. Rev. C **43**, 849 (1991)
27. S. Okada et al: Nucl. Phys. A **754**, 178c (2005)
28. J. D. Bjorken and S. D. Drell: *Relativistic Quantum Mechanics* (McGraw-Hill, N.Y., 1964)
29. B. H. J. McKellar and B. F. Gibson: Phys. Rev. C **30**, 322 (1984)
30. K. Takeuchi, H. Bandō, and H. Takaki: Prog. Theor. Phys. **73**, 841 (1985)

31. L. de la Torre: Ph.D. Thesis, Univ. of Massachusetts (1982)
32. A. Parreño: Ph.D. Thesis, Univ. of Barcelona (1997); http://www.ecm.ub.es/assum/tesi.ps
33. J. F. Donoghue et al: Phys. Rep. **131**, 319 (1986)
34. M. Shmatikov: Nucl Phys. A **580**, 538 (1994)
35. K. Itonaga, T. Ueda, and T. Motoba: Phys. Rev. C **65**, 034617 (2002)
36. D. Jido, E. Oset, and J. E. Palomar: Nucl. Phys. A **694**, 525 (2001)
37. T. Inoue, S. Takeuchi, and M. Oka: Nucl. Phys. A **597**, 563 (1996)
38. M. K. Gaillard and B. W. Lee: Phys. Rev. D **10**, 897 (1974); G. A. Altarelli and L. Maiani: Phys. Lett. B **52**, 351 (1974); F. J. Gilman and M. B. Wise: Phys. Rev. D **20**, 2392 (1979); **27**, 1128 (1983)
39. E. A. Paschos, T. Schneider, and Y. L. Wu: Nucl. Phys. B **332**, 285 (1990); G. Buchalla, A. J. Buras, and M. K. Harlander: Nucl. Phys. B **337**, 313 (1990)
40. A. Parreño, A. Ramos, C. Bennhold, and K. Maltman: Phys. Lett. B **435**, 1 (1998)
41. J.-H. Jun and H. C. Bhang: Nuovo Cim. A **112**, 649 (1999); J.-H. Jun, Phys. Rev. C **63**, 044012 (2001)
42. A. Parreño, C. Bennhold, and B. R. Holstein: Nucl. Phys. A **754**, 127c (2005); A. Parreño, C. Bennhold, and B. R. Holstein: Phys. Rev. C **70**, 051601 (2004)
43. R. H. Dalitz: In Proc. *Summer Study Meeting on Nuclear and Hypernuclear Physics with Kaon Beams*, BNL Report No. 18335 (1973) p. 41
44. C. -Y. Cheung, D. P. Heddle, and L. S. Kisslinger: Phys. Rev. C **27**, 335 (1983); D. P. Heddle and L. S. Kisslinger: Phys. Rev. C **33**, 608 (1986)
45. K. Sasaki, T. Inoue, and M. Oka: Nucl. Phys. A **669**, 331 (2000); (E) **678**, 455 (2000); **707**, 477 (2002)
46. H. Noumi et al: Phys. Rev. C **52**, 2936 (1995)
47. H. Noumi et al: In Proc. *IV International Symposium on Weak and Electromagnetic Interactions in Nuclei*, ed by H. Ejiri, T. Kishimoto, and T. Sato (World Scientific, 1995) p. 550
48. H. C. Bhang et al: Phys. Rev. Lett. **81**, 4321 (1998); H. Park et al: Phys. Rev. C **61**, 054004 (2000)
49. H. Outa et al: Nucl. Phys. A **670**, 281c (2000)
50. B. H. Kang et al: Phys. Rev. Lett. **96**, 062301 (2006)
51. M. J. Kim et al: Phys. Lett. B **641**, 28 (2006)
52. K. Miyagawa and W. Glöckle: Phys. Rev. C **48**, 2576 (1993); J. Golak et al: Phys. Rev. C **55**, 2196 (1997); (E)**56**, 2892 (1997); J. Golak et al: Phys. Rev. Lett. **83**, 3142 (1999)
53. V. G. J. Stoks and T. A. Rijken: Phys. Rev. C **59**, 3009 (1999); T. A. Rijken, V. G. J. Stoks, and Y. Yamamoto: Phys. Rev. C **59**, 21 (1999)
54. E. Hiyama, Y. Kino, and M. Kamimura: Prog. Part. Nucl. Phys. **51**, 223 (2003)
55. D. Halderson: Phys. Rev. C **48**, 581 (1993)
56. M. M. Nagels, T. A. Rijken, and J. J. de Swart: Phys. Rev. D **15**, 2547 (1977); P. M. M. Maessen, T. A. Rijken, and J. J. de Swart: Phys. Rev. C **40**, 2226 (1989)
57. B. Holzenkamp, K. Holinde, and J. Speth: Nucl. Phys. A **500**, 485 (1989)
58. J. Haidenbauer and U.-G. Meissner: Phys. Rev. C **72**, 044005 (2005)
59. A. Ramos, M. J. Vicente-Vacas, and E. Oset: Phys. Rev. C **55**, 735 (1997); (E) **66**, 039903 (2002)
60. W. M. Alberico, A. De Pace, M. Ericson, and A. Molinari: Phys. Lett. B **256**, 134 (1991)

61. A. Ramos, E. Oset, and L. L. Salcedo: Phys. Rev. C **50**, 2314 (1994)
62. E. Bauer and F. Krmpotić: Nucl. Phys. A **739**, 109 (2004)
63. J. H. Kim el al: Phys. Rev. C **68**, 065201 (2003)
64. S. Ajimura et al: Phys. Lett. B **282**, 293 (1992)
65. A. Ramos, E. van Meijgaard, C. Bennhold, and B. K. Jennings: Nucl. Phys. A **544**, 703 (1992)
66. W. M. Alberico, G. Garbarino, A. Parreño, and A. Ramos: Phys. Rev. Lett. **94**, 082501 (2005)
67. T. Maruta et al: Nucl. Phys. A **754**, 168c (2005)
68. C. Ordonez, L. Ray, and U. van Kolck: Phys. Rev. C **53**, 2086 (1996), in particular Sect. V; U. van Kolck: Prog. Part. Nucl. Phys. **43**, 337 (1999); S. R. Beane et al: nucl-th/0008064 (Boris Ioffe Festschrift, ed. by M. Shifman, World Scientific)
69. This approach to chiral SU(3) has, in particular, been advocated by J. F. Donoghue and B. R. Holstein: Phys. Lett. B **436**, 331 (1998) and J. F. Donoghue, B. R. Holstein, and B. Borasoy: Phys. Rev. D **59**, 036002 (1999)
70. O. Hashimoto et al: Phys. Rev. Lett. **88**, 042503 (2002)
71. S. Ajimura et al: Phys. Rev. Lett. **84**, 4052 (2000)
72. O. Hashimoto et al: Erratum of [70] (to be submitted); Y. Sato: poster presented at PANIC02, Osaka (Japan), Sept 30–Oct 4, 2002; nucl-ex/0409007
73. C. Barbero and A. Mariano: Phys. Rev. C **73**, 024309 (2006)
74. K. Sasaki, M. Izaki, and M. Oka: Phys. Rev. C **71**, 035502 (2005)
75. This point was addressed by the seminar given by C. Chumillas in the School
76. S. R. Beane, P. F. Bedaque, A. Parreño, and M. J. Savage: Nucl. Phys. A **747**, 55 (2005)

Index